"This is an excellent, well-organized, clearly written, and comprehensive book. It does a great job both with expounding the details of Cassirer's work, at a level that will be appropriate for advanced undergraduate and graduate students, and showing the deep and lasting relevance of Cassirer's thought today."

—Paul Livingston, University of New Mexico, USA

"Matherne has written a guide to Cassirer's philosophy that is both accurate and stimulating to read. She takes into account the whole scope of his work - from his early writings and his professorship during the Weimar Republic, to his years of exile in Sweden and the United States - and provides an important addition to our understanding of his lasting influence."

—Jeffrey Andrew Barash, Université de Picardie, Amiens, France

Cassirer

Ernst Cassirer (1874–1945) occupies a unique place in 20th-century philosophy. His view that human beings are not rational but symbolic animals and his famous dispute with Martin Heidegger at Davos in 1929 are compelling alternatives to the deadlock between 'analytic' and 'continental' approaches to philosophy. An astonishing polymath, Cassirer's work pays equal attention to mathematics and natural science but also art, language, myth, religion, technology, and history. However, until now the importance of his work has largely been overlooked.

In this outstanding introduction Samantha Matherne examines and assesses the full span of Cassirer's work. Beginning with an overview of his life and works she covers the following important topics:

- Cassirer's Neo-Kantian background
- Philosophy of mathematics and natural science, including Cassirer's first systematic work, Substance and Function, and subsequent works, like Einstein's Theory of Relativity
- The problem of culture and the ground-breaking The Philosophy of Symbolic Forms
- Cassirer's ethical and political thought and his diagnosis of fascism in The Myth of the State
- Cassirer's influence and legacy.

Including chapter summaries, suggestions for further reading, and a glossary of terms, this is an ideal introduction to Cassirer's thought for anyone coming to his work for the first time. It is essential reading for students in philosophy as well as related disciplines such as intellectual history, art history, politics, and literature.

Samantha Matherne is an Assistant Professor of Philosophy at Harvard University, USA. She has published articles on Kant, Neo-Kantianism, phenomenology, and aesthetics. She is a co-author (with Dominic McIver Lopes, Mohan Matthen, and Bence Nanay) of the forthcoming book *The Geography of Taste*.

Samantha Matherne

Cassirer

Routledge
Taylor & Francis Group

LONDON AND NEW YORK

First published 2021
by Routledge
2 Park Square, Milton Park, Abingdon, Oxon OX14 4RN

and by Routledge
52 Vanderbilt Avenue, New York, NY 10017

Routledge is an imprint of the Taylor & Francis Group, an informa business

© 2021 Samantha Matherne

British Library Cataloguing-in-Publication Data
A catalogue record for this book is available from the British Library

Library of Congress Cataloging-in-Publication Data
Names: Matherne, Samantha, author.
Title: Cassirer / Samantha Matherne.
Description: Abingdon, Oxon; New York: Routledge, 2021. |
   Includes bibliographical references and index. |
   Identifiers: LCCN 2020048569 (print) | LCCN 2020048570 (ebook)
Subjects: LCSH: Cassirer, Ernst, 1874–1945.
Classification: LCC B3216.C34 M375 2021 (print) |
   LCC B3216.C34 (ebook) |DDC 193—dc23
LC record available at https://lccn.loc.gov/2020048569
LC ebook record available at https://lccn.loc.gov/2020048570

ISBN: 978-1-138-82749-3 (hbk)
ISBN: 978-1-138-82750-9 (pbk)
ISBN: 978-1-351-04885-9 (ebk)

Typeset in Joanna
by Apex CoVantage, LLC

For my parents,
Roman and Elizabeth Matherne

Contents

Acknowledgments

This project dates back to my time in graduate school at the University of California, Riverside, where my initial interest in Cassirer was sparked, thanks to the encouragement of my advisor, Pierre Keller, and our reading groups on Cassirer in Erich Reck's living room. The opportunity to develop these glimmers into a book is something I owe to Brian Leiter and Tony Bruce, who initially approached me about this volume. This manuscript could not have been written without the institutions and people who have since seen me through.

At the institutional level, I would like to thank the students, faculty, departments of philosophy, and support-structures in place at my various home institutions during this time: the University of British Columbia; the University of California, Santa Cruz; and Harvard University. I am also grateful to the National Endowment for the Humanities for awarding me a Fellowship for University Teachers for this project in 2016–17. My research also benefitted from my time with Cassirer's papers at Yale's Beinecke Library and the Warburg Institute in London.

I am indebted to many people for feedback on various parts of this manuscript, including Fabien Capeillères, Tobias Endres, Steve Lofts, Roman J. Matherne, Jr., Anne Pollok, Erich Reck, Tom Ryckman, and three anonymous referees.

On a more personal note, at every stage along the way I have been buoyed by the encouragement and cheer of so many. Many 'thank yous' are in order. To Scott Edgar, Lydia Patton, and Alan Richardson: for your distinctive brand of Neo-Kantian solidarity.

To Peter Gordon: for your distinctive brand of Post-Kantian solidarity. To David Landy: for reading Kant across the Bay Area. To Nico Orlandi and Janette Dinishak: for accompanying me through the redwoods and banana slugs. To Susanna Siegel: for the cafés and camaraderie. To Crista Farrell and Mark Leher: for being my oldest philosophical interlocutors. To Vilashini Cooppan, Jennifer Derr, Jody Green, Marc Matera, and the rest of the Santa Cruz Lot: for the lifting of spirits, in all senses. To Christine Catania: for the pleasure of all of the *italienischen Reisen*. To Emmanuelle Humblet: for breaking it down since 2001. To Aimee Baker: for your humor and hope, and the happiness that is Carolena and Carlos. Finally, and most of all, to my parents: for supporting the writer in me since I could hold a crayon.

Abbreviations

AH   *Axel Hägerström*
References are to ECW 21.

AS   "Albert Schweitzer as Critic of Nineteenth-Century Ethics"
References are to ECW 24.

CIPC   "Critical Idealism as a Philosophy of Culture"
References are to *Symbol, Myth, and Culture*/pagination of ECN 7.

CP   "The Concept of Philosophy as a Philosophical Problem"
References are to *Symbol, Myth, and Culture*/pagination of ECN 9.

CSF   "The Concept of Symbolic Form in the Construction of the Human Sciences"
References are to *The Warburg Years*/pagination of ECW 16.

Corr.   Correspondence in *Ausgewählter wissenschaftlicher Briefwechsel*.
References are to ECN 18.

DI   *Determinism and Indeterminism in Modern Physics*
References are to the Benfey translation/pagination of ECW 19.

ECN   *Nachgelassene Manuskripte und Texte*

ECW   *Gesammelte Werke: Hamburger Ausgabe*

EM   *An Essay on Man*
References are to the Yale edition/pagination of ECW 23.

ETR   *Einstein's Theory of Relativity*
References are to the Swabey and Swabey translation/pagination of ECW 10.

EVA   "The Educational Value of Art"
References are to *Symbol, Myth, and Culture*.

FT   "Form and Technology"
References are to The Warburg Years/pagination of ECW 17.

IRC   "The Idea of a Republican Constitution"
References are to the Berk translation/pagination of ECW 17.

KLT   Kant's Life and Thought
References are to the Haden translation/pagination of ECW 8.

KMM   "Kant and Modern Mathematics" ("Kant und die moderne Mathematik")
References are to ECW 9/pagination of Kant Studien.

LCS   The Logic of the Cultural Sciences
References are to the Lofts translation/pagination of ECW 24.

LCW   "Language and the Construction of the World of Objects"
References are to The Warburg Years/ECW 18.

LM   "Language and Myth"
References are to The Warburg Years/ECW 16.

MATS   "Mythic, Aesthetic, and Theoretical Space"
References are to The Warburg Years/ECW 17.

MS   The Myth of the State
References are to the Yale edition/ECW 25.

MSF   "On the Metaphysics of Symbolic Forms."
References are to The Philosophy of Symbolic Forms, Vol. 4./ECN 1.

NDNR   "On the Nature and Development of Natural Right" ("Vom Wesen und Werden des Naturrechts")
References are to ECW 18.

PK   The Problem of Knowledge: Philosophy, Science, and History Since Hegel
References are to the Woglom and Hendel translation/pagination of ECW 5.

PP   "Philosophy and Politics"
References are to Symbol, Myth, and Culture.

PSF   "The Problem of the Symbol"
References are to The Warburg Years/ECW 17.

PSF   The Philosophy of Symbolic Forms
References to each volume are to the Lofts translations/pagination of ECW 11 (v1), 12 (v2), 13 (v3)/pagination of the Mannheim translations.

SF   *Substance and Function*
            References are to the Swabey and Swabey translation/
            pagination of ECW 6.
TMPM   "The Technique of Our Modern Political Myths"
            References are to *Symbol, Myth, and Culture.*

Chronology

1874 Born in Breslau, Silesia (now Wroclaw, Poland), to Eduard and Eugenie (Jenny), July 28

1892 Enters Friedrich Wilhelm University (now the Humboldt University of Berlin)

1896 Moves to the University of Marburg to study with Hermann Cohen

1899 Publishes his inaugural dissertation, *Descartes's Critique of Mathematical and Natural-Scientific Cognition* (*Descartes' Kritik der mathematischen und naturwissenschaftlichen Erkenntnis*)

1901 Publishes *Leibniz's System in Its Scientific Foundations* (*Leibniz' System in seinen wissenschaftlichen Grundlagen*)
Marries Antonelle (Toni) Bondy

1903 Moves to Berlin

1906 Publishes the first volume of *The Problem of Cognition in Philosophy and Science in the Modern Age* (*Das Erkenntnisproblem in der Philosophie und Wissenschaft der neueren Zeit*), which earns him his habilitation
Appointed as a *Privatdozent* at the Friedrich Wilhelm University

1907 Publishes the second volume of *The Problem of Cognition in Philosophy and Science in the Modern Age* (*Das Erkenntnisproblem in der Philosophie und Wissenschaft der neueren Zeit*).

1910 Publishes *Substance and Function*

1914 Wins Kuno-Fischer Gold Medal for *The Problem of Cognition* (*Das Erkenntnisproblem*)
Appointed to the Press Office in Berlin for his civil service during World War I

1919   Appointed as a full professor at the newly founded University
       of Hamburg
1920   Publishes the third volume of The Problem of Cognition in the
       Modern Age: The Post-Kantian System (Das Erkenntnisproblem in der
       Philosophie und Wissenschaft der neueren Zeit. Die nachkantischen System)
       Introduced to the Warburg Library for Cultural Sciences
       (now Warburg Institute in London)
1921   Publishes Einstein's Theory of Relativity
1923   Publishes The Philosophy of Symbolic Forms, Volume One: Language
1925   Publishes The Philosophy of Symbolic Forms, Volume Two: Mythical
       Thinking
1929   Publishes The Philosophy of Symbolic Forms, Volume Three: The
       Phenomenology of Cognition
       Davos disputation with Heidegger
       Elected as the Rector of the University of Hamburg
1932   Publishes The Philosophy of the Enlightenment
1933   Leaves Germany after Hitler becomes Chancellor
       Accepts position at All Souls College at Oxford
1935   Accepts position at the University of Göteborg in Sweden
1936   Publishes Determinism and Indeterminism in Modern Physics
1939   Becomes a Swedish citizen
1941   Accepts visiting position at Yale
1942   Publishes The Logic of the Cultural Sciences
1944   Accepts visiting position at Columbia
       Publishes An Essay on Man
1945   Dies of a heart attack in New York City, April 13
1946   The Myth of the State is published posthumously.

One
Cassirer's life and works

Introduction

Ernst Cassirer is best known as a leading figure in the German Neo-Kantian movement, and, though he is this, his philosophical system belongs just as much to a longer-standing humanist tradition that pursues the question, 'what is it to be a human being?' Cassirer answers: it is to be someone who speaks a language, experiences art, and explores science. It is to be someone who is shaped as much by myth and religion, as by morality and politics. It is to be someone who is situated in history, who thinks mathematically, and who engages with technology. In short, to be a human being is to be someone who participates in a shared cultural world.

In developing this culturally-situated account of the human being, Cassirer built one of the most detailed and comprehensive philosophical systems of the 20th century. He made seminal contributions not only to the philosophical study of mathematics and the natural sciences (*Naturwissenschaften*), but also to aesthetics, history, moral-political thought, the philosophy of language, and other issues in the so-called 'human sciences' (*Geisteswissenschaften*). Indeed, few other bodies of philosophical work in the 20th century rival his in scope. What is more, in each of these endeavors, Cassirer was guided by a commitment to do philosophy not 'from the armchair', but through the work of interdisciplinary engagement. He thus arrived at his positions as the result of careful study not just of philosophy, but of historical and contemporary research in fields outside of philosophy.

In a century in which the divisions within philosophy and across the academy became more entrenched, the synthetic nature of Cassirer's thought stands out. His is a thought capable of bridging gaps not only between the so-called 'analytic-continental divide' in philosophy, but also between philosophy and other disciplines in the humanities, social sciences, and natural sciences.

Though the scope of Cassirer's work is a source of its power, it also serves as one of the greatest challenges facing his readers. Indeed, Cassirer not only engages in a wide range of topics, he also develops a style that is multifaceted. He typically pairs philosophical analysis with a wealth of details drawn from research in other disciplines. Though impressive in its erudition, this style of writing can make it difficult to pinpoint exactly what Cassirer's own philosophical views are. In light of this challenge, my goal in this book is to offer an overview of Cassirer's philosophical system as a whole that can help readers navigate his corpus.

In order to explore his system, I largely follow Cassirer's own intellectual trajectory, beginning with his early work on mathematics and natural science, before turning to the philosophy of culture that he develops later in his career.[1] However, my aim throughout is systematic in nature: I endeavor to bring to light the philosophical system that Cassirer develops through this body of work. For this reason, I set the stage with an analysis of the basic Neo-Kantian commitments that lay the groundwork for his philosophical system. And in the discussion that ensues, I show how this systematic framework shapes his philosophical treatment of mathematics, natural science, and culture.

In the rest of this chapter, I offer an overview of Cassirer's life and works. I then turn to the Neo-Kantian framework that underwrites his philosophy as a whole in Chapter Two. In Chapters Three and Four I shift my attention to Cassirer's philosophy of mathematics and natural science, respectively. Next in Chapters Five, Six, and Seven, I investigate Cassirer's philosophy of culture. I begin with a discussion of Cassirer's systematic approach to culture in Chapter Five. In Chapters Six and Seven I consider his account of the specific regions ('symbolic forms') of culture, including myth, religion, art, language, history, technology, mathematics, natural science, and right (*Recht*). In Chapter Seven, I pay particular attention to the ethical and political dimensions of his philosophy of culture. I conclude in Chapter Eight with remarks about Cassirer's legacy in the 20th and 21st centuries.

As for my purposes in this chapter, I divide my overview of Cassirer's life and works into four sections: the early years, the Berlin years, the Hamburg years, and the years in exile.

The early years (1874–1902)

Cassirer was born on July 28, 1874 in Breslau, Silesia (now Wroclaw, Poland) into an affluent, cosmopolitan, Jewish family. The son of a merchant, Eduard (1843–1916), and his wife Eugenie (Jenny) (1848–1904), Cassirer was the third of seven children and he spent much of his childhood visiting extended family in Berlin. Among Cassirer's cousins were Bruno Cassirer (1872–1941), the renowned publisher; Kurt Goldstein (1878–1965), the famous neurologist and psychiatrist; Richard Cassirer (1868–1925), another neurologist; Fritz Cassirer (1871–1926), the conductor; and Paul Cassirer (1871–1926), the art dealer who helped introduce the art of Cézanne, van Gogh, and other Impressionists and Post-Impressionists to Germany.

In 1892 Cassirer entered the Friedrich Wilhelm University (now the Humboldt University of Berlin) with plans to study German literature and philosophy. As was common at that time, Cassirer took courses not just at his home institution in Berlin, but at universities in Leipzig, Heidelberg, and Munich as well. However, his education took a decisive turn in the summer of 1894 when he, back in Berlin, took a course on Kant with Georg Simmel (1858–1915). Of particular significance was Simmel's remark that, "Undoubtedly the best books on Kant are written by Hermann Cohen; but I must confess that I do not understand them" (Gawronsky 1949: 6). Hearing Simmel's praise, Cassirer went to the bookstore, ordered all of Cohen's available volumes, and quickly resolved to move to the University of Marburg to pursue his graduate work with Cohen.

Hermann Cohen (1842–1918) was one of the most important German Jewish philosophers of the 19th and early 20th centuries.[2] He was a key figure in the Neo-Kantian movement that dominated academic philosophy in Germany from 1870 to 1920 (see Chapter Two). And he founded one of the leading schools of Neo-Kantianism, the so-called 'Marburg School', whose most famous students include Paul Natorp (1854–1924) and Cassirer. Cohen also made seminal contributions to Jewish thought, both in his academic work and as a public intellectual.

Once Cassirer arrived at Marburg in 1896, he immediately stood out among his contemporaries. Indeed, after Cassirer's first seminar with Cohen, Cohen remarked, "I felt at once that this man had nothing to learn from me" (Gawronsky 1949: 7). And though during these years, he developed a close relationship with Cohen and Natorp, he also spent much of his time absorbed in his studies, behavior which earned him the lifelong nickname, 'the Olympian'.[3]

Taking his cue from Cohen and Natorp, Cassirer conducted his early research in the field of *Erkenntnistheorie*, which can be translated, more literally, as 'theory of cognition' or 'theory of knowledge', or, more loosely, as 'epistemology'.[4] In keeping with this Marburg trend, Cassirer wrote a dissertation on Descartes, titled *Descartes's Critique of Mathematical and Natural-Scientific Cognition* (*Descartes' Kritik der mathematischen und naturwissenschaftlichen Erkenntnis*) (1899), which earned the prestigious 'opus eximium'. Cassirer followed up his dissertation three years later with *Leibniz's System in Its Scientific Foundations* (*Leibniz' System in seinen wissenschaftlichen Grundlagen*) (1902), the first part of which was his treatise on Descartes and the second part of which was a systematic analysis of Leibniz's philosophy, which addressed Leibniz's mathematics, mechanics, metaphysics, ethics, philosophy of right, aesthetics, and theodicy. This work earned Cassirer the second prize from the Berlin Academy (no first prize was given).

The publication of *Leibniz's System* was not the only significant event for Cassirer in 1902: that year he married Antonelle (Toni) Bondy (1883–1961) from Vienna. And together, the couple had three children: Heinrich (Heinz) (1903–1979), who would later become a Kant scholar and student of H.J. Paton; Georg (1904–1958), who became a photographer; and Anna (1908–1998), who became a psychotherapist.

After leaving Marburg, Cassirer spent a year in Munich, during which time he seriously debated whether to remain in academia given concerns over the anti-Semitism prevalent in many of the smaller university towns in which he was likely to find a job. Resolving for academia in the end, in 1903 Cassirer returned to Berlin.

The Berlin years (1903–1919)

During his early years in Berlin, Cassirer did not hold any academic post, devoting his energy instead to writing, what would become, two of his most important works: volumes one and two of The Problem of Cognition in Philosophy and Science in Modern Times (Das Erkenntnisproblem in der Philosophie und Wissenschaft der neueren Zeit) (1906, 1907). In these expansive volumes in the history of philosophy Cassirer traces the development of the concept of cognition in philosophy, natural science, and the humanities from the Renaissance to Kant.

Cassirer submitted the first volume of The Problem of Cognition for his habilitation, which also required that he give a public address. The topic of Cassirer's lecture was the Kantian notion of a 'thing in itself' (Ding an sich) and it drew sharp criticism from two leading philosophers, Carl Stumpf (1848–1936) and Alois Riehl (1844–1924).[5] However, Wilhelm Dilthey (1833–1911) came to his defense, claiming that, "I would not like to be a man of whom posterity will say that he rejected Cassirer" (Gawronsky 1949: 17). Ultimately it was enough and Cassirer earned his habilitation and accepted a Privatdozent (instructor) position at the Friedrich Wilhelm University, where he would remain until 1919.

In his first years as a Privatdozent, Cassirer focused his attention on providing a Neo-Kantian account of the developments in modern mathematics and natural science, specifically in the fields of physics and chemistry. To this end, in "Kant and Modern Mathematics" ("Kant und die moderne Mathematik") (1907), Cassirer argued that in order to account for 19th century developments in geometry and number theory, a Kantian approach is needed rather than the logicist approach of Bertrand Russell (1872–1970) and Louis Couturat (1868–1914). This article, in turn, set up what would be Cassirer's major achievement during this period, Substance and Function: Investigations of the Fundamental Questions of the Critique of Cognition (Substanzbegriff und Funktionsbegriff: Untersuchungen über die Grundfragen der Erkenntniskritik) (1910). Cassirer himself describes Substance and Function as his first 'systematic', rather than historical, monograph. And in it he defends a Neo-Kantian approach to modern mathematics, classical mechanics, and chemistry (PSFv3 xxxii/vii/xiii).

After *Substance and Function*, Cassirer turned his attention to a number of different projects, including editing a new volume of Kant's collected works, which was published by his cousin, Bruno Cassirer, and writing an overview of Kant's philosophical development to accompany it, titled *Kant's Life and Thought* (*Kants Leben und Lehre*) (1918). In this period, he also published *Freedom and Form* (*Freiheit und Form*) (1916), in which he offers an analysis of how the notions of 'freedom' and 'form' develop in the tradition of German liberalism in the 18th–19th centuries. To this end, he focuses on broad developments in aesthetics and political theory, as well as on the work of Leibniz, Kant, Goethe, Schiller, and Fichte. In addition to these scholarly activities and teaching responsibilities, during World War I Cassirer did his civil service in the Press Office in Berlin, where his job was to report on international news on the war.

Yet in spite of this prodigious output, Cassirer was never offered a full professorship at the Friedrich Wilhelm University. Indeed, he was not appointed as a full professor until 1919, when he accepted a position at the newly founded University of Hamburg.

The Hamburg years (1919–1933)

Cassirer's time in Hamburg was among the most fruitful periods of his life and it was during these years that he developed a systematic philosophy of culture. This project was aided in no small part by his participation at the Warburg Library for Cultural Sciences (*Kulturwissenschaftliche Bibliothek Warburg*) in Hamburg (now the Warburg Institute in London). Aby Warburg (1866–1929) was a member of a wealthy family of Jewish bankers in Hamburg, who worked as an independent scholar, specializing in Renaissance art history and cultural theory. Over the course of his career, Warburg built a vast personal library, with primary and secondary texts on myth, language, art, religion, philosophy, and science. And in the early 1920s, thanks to the efforts of the art historians Fritz Saxl (1890–1944) and Gertrud Bing (1892–1964) (who wrote her dissertation on Leibniz and Lessing under Cassirer), the library became a leading interdisciplinary research institution. It drew prominent scholars, like the art historians Erwin Panofsky (1892–1968), Edgar Wind (1900–1971), and Gustav Pauli (1866–1938); historians, like Paul Oskar Kristeller (1905–1999) and Richard Salomon (1884–1966); the Middle East scholar, Hellmut Ritter (1892–1971); the philologist, Karl Reinhardt (1886–1958), among others. Though it had become affiliated with the University of Hamburg, with Hitler's rise to power in 1933, the Library was relocated to London and reopened, still under the guidance of Saxl and Bing, as the Warburg Institute.

Upon entering the Warburg Library for the first time, Cassirer remarked to Saxl,

> This library is dangerous. I shall either have to avoid it altogether or imprison myself here for years. The philosophical problems involved are close to my own, but the concrete historical material which Warburg has collected is overwhelming.
>
> (Saxl 1949: 48)

The "philosophical problems" Cassirer has in mind are problems related to developing a systematic approach to culture, which had begun to occupy him in Berlin.[6] And in addition to the "concrete historical material" that Cassirer used for research, the Warburg Library offered him an interdisciplinary community, which he quickly became an integral part of.

It was in this stimulating environment that Cassirer wrote his three-volume masterwork, The Philosophy of Symbolic Forms (Philosophie der symbolischen Formen). Although over the course of these volumes Cassirer develops a systematic philosophy of culture, in each volume he focuses on a more specific region, or "symbolic form", of culture. In the first volume, Language (1923), he singles out the symbolic form of language. In the second volume, Mythical Thinking (1925), he analyzes the symbolic forms of myth and religion. And in the third volume, Phenomenology of Cognition (1929), he considers how the symbolic forms of mathematics and natural science develop out of the other symbolic forms. In addition to these volumes, during this period Cassirer published a series of articles on culture, many of which were presented in some form at the Warburg Library, including "The Concept of Symbolic Form in the Construction of the Human Sciences" ("Der Begriff der Symbolischen Form im Aufbau der Geisteswissenschaften") (1923), "Language and Myth" ("Sprache und Mythos") (1925), "The Problem of the Symbol and Its Place in the System of Philosophy" ("Das Symbolproblem und Seine Stellung im System der Philosophie") (1927), "Form and Technology" ("Form und Technik") (1930), among others (many of which have been collected in The Warburg Years, 2013). In 1928 Cassirer also began writing a manuscript for a fourth volume of The Philosophy of Symbolic Forms, tentatively titled, "The Metaphysics of Symbolic Forms" ("Zur Metaphysik der symbolischen Formen"), in which he compares his account of the human being and culture with the approaches endorsed by thinkers in the traditions of Lebensphilosophie and philosophical anthropology, like Schopenhauer, Nietzsche, Kierkegaard, Dilthey, Bergson, and Simmel. This manuscript was posthumously published in English as The Philosophy of Symbolic Forms, Volume 4: The Metaphysics of Symbolic Forms (1996), alongside other related manuscripts, including "Symbolic Forms: For Volume Four" ("Symbolische Formen. Zu Band 4") (1928), "Appendix: The Concept of the Symbol: Metaphysics of the Symbolic" ("Beilage: Symbolbegriff: Metaphysik des Symbolischen") (c. 1928), and "On Basis Phenomena" ("Über Basisphänomene") (c. 1940). In the latter manuscript, Cassirer offers a Goethe-inspired discussion of the "basis phenomena" of "I", "you", and "work".[7]

However, Cassirer's output in this period was not confined to the philosophy of culture; he continued to develop his earlier work in the history of philosophy, philosophy of science, and political thought. In the historical vein, he published the third volume of The Problem of Cognition: The Post-Kantian System (Das Erkenntnisproblem in der Philosophie und Wissenschaft der neueren Zeit. Die nachkantischen System) (1920), dedicated to post-Kantian developments in the long 19th century, including Kant's early reception by Reinhold, Beck, and Maimon; the German Idealism of Fichte, Schelling, and Hegel; and the alternatives to German Idealism proposed by Schopenhauer, Herbart, and Fries. He also wrote several landmark texts on the philosophy of the Renaissance and Enlightenment, including Idea and Gestalt: Goethe, Schiller, Hölderlin, Kleist (Idee und Gestalt. Goethe, Schiller, Hölderlin, Kleist) (1921), The Individual and the Cosmos (Individuum und Kosmos in der Philosophie der Renaissance) (1927), The Philosophy of Enlightenment (Die Philosophie der *Aufklärung*) (1932), The *Platonic Renaissance in England* (*Die Platonische Renaissance in England und die Schule von Cambridge*) (1932), *Goethe and the Historical World* (*Goethe und die geschichtliche Welt*) (1932), and *The Question of Jean-Jacques Rousseau* ("*Das Problem Jean Jacques Rousseau*") (1932–33). The flurry of texts around 1932, in particular, were inspired by Cassirer's time in Paris at the National Library of France (*Bibliothèque nationale de France*), where he had been invited to spend the summer of 1931.

In addition to these historically-oriented writings, Cassirer wrote *Einstein's Theory of Relativity* (*Zur Einsteinschen Relativitätstheorie*) (1921) and became one of the first philosophers to engage with relativity. Meanwhile, in a political vein, taking up some of the themes in political liberalism that he had begun addressing in *Freedom and Form*, Cassirer produced political texts during this period: "The Idea of the Republican Constitution" ("*Die Idee der republikanischen Verfassung*") (1928), which he presented at a celebration of the 10th anniversary of the Weimar Republic, and "On the Nature and Development of Natural Right" ("*Vom Wesen und Werden des Naturrechts*") (1932).

Of Cassirer's years in Hamburg, 1929 was particularly significant. In July he was elected as the Rector of the University of Hamburg. And, earlier in the spring, he had the now famous disputation with Martin Heidegger (1889–1976) in Davos, Switzerland.[8] The disputation between Cassirer and Heidegger took place as part of a week-long International University Course (*Hochschulkurse*), which drew intellectuals from a wide range of fields, including Rudolph Carnap (1891–1970), Emmanuel Levinas (1906–1995), Eugen Fink (1905–1975), Erich Maria Remarque (1898–1970), among others. The guiding question for the course was, "What is the Human Being?," which Cassirer and Heidegger addressed both in independent lectures and in a public debate. In retrospect, many have read underlying political tensions into this debate, with Cassirer as the older, liberal Jew pitted against Heidegger, as the younger thinker with National Socialist leanings. However, the actual debate between Cassirer and Heidegger was reported to have been quite collegial with the major disagreements being philosophical, rather than political.

This said, the philosophical disagreement was substantive. Each responded to the question, "What is the Human Being?," in opposing terms. For his part, Cassirer defended a more Kantian account of the human being that emphasized freedom and spontaneity, whereas Heidegger defended a more existential conception of the human that emphasized our 'thrownness' (Geworfenheit) into the world. Kant, too, was a point of contention, as Cassirer and Heidegger clashed with regard to whether Kant ought to be read along the more epistemological (erkenntnistheoretisch) lines favored by the Neo-Kantians or the more metaphysical lines favored by Heidegger. This latter debate continued to play out after Davos with Heidegger's publication of Kant and the Problem of Metaphysics (1929 [1990]) and Cassirer's critical review of it in Kant-Studien in 1931.

Cassirer and Heidegger met one more time in Freiburg in 1932, where Heidegger helped secure Cassirer an invitation to lecture. However, with Hitler's rise to power in 1933, their paths quickly diverged: while Heidegger joined the Nazi party and was appointed the Rector of Freiburg, Cassirer resigned from Hamburg and took his family into exile.

Cassirer himself offers a description of what his resignation from Hamburg meant to him in a letter to the Dean of the University of Hamburg, Walther Küchler, in 1933. Responding to reports in the Frankfurter Zeitung that he had taken leave "for private reasons" and the Hamburger Fremdenblatt that he had done so for "for health concerns," Cassirer writes,

> In the early days of April, immediately after the first news of the boycott movement against the German Jews, I sent a letter to the university authorities, as well as to the Rector of the University, in which I explained in detail the reasons why under the present circumstances it was, unfortunately, no longer possible for me to hold my position as a university professor in Germany. These grounds were in no way 'private'; they were of a purely principled nature. I think too highly of the significance and dignity of the academic position to be able to hold it at a time in which, to me as a Jew, collaboration in German cultural-work [deutschen Kulturarbeit] has been denied, or diminished and reduced through legal means. The work I have been able to do so far within the faculty was based on the fact that I was recognized as an equal member; and only under this condition did that work receive its sense [Sinn] and content [Inhalt]. With the elimination of this

condition, every possibility for me to join in the work of the faculty in a substantive, fruitful way disappeared. Therefore in my petition, which was made *before* the emergence of the new civil service—I expressly asked not for leave from my lectures and exercises, but for release from all official duties. . . . Thus . . . I must regard the tie that bound me up until now to the Philosophy Faculty at the University of Hamburg severed. What this dissolution [*Lösung*] means for me: there are no words for it. But the one thing I can say is that in all the deep grief over the events of the last weeks and the fate of the German youth, the feeling of the inner connection [*Verbundenheit*] with the goals [*Aufgaben*] and fortunes [*Geschicken*] of the University of Hamburg has not left me. Whatever my work and my personal fortune may be from now on, the years of my activity at the University of Hamburg, whose highest academic office I was able to hold, will not be forgotten or lost to me.

(Warburg Archive General Correspondence, 1933/1643,
E. Cassirer to W. Küchler, 27 April 1933, my transl.)

Cassirer would never hold an academic post in Germany again, spending his remaining years in exile.

Years in exile (1933–1945)

After leaving Hamburg, Cassirer spent two years at Oxford (1933–1935), six years at the University of Göteborg in Sweden, (1935–1941), and four years in the United States, first at Yale and then at Columbia (1941–1945). And though Cassirer had only a reading knowledge of English and did not know Swedish at all, he quickly became fluent in both and was able to flourish in the new intellectual communities in which he found himself.

During this period, Cassirer carried forward many of his long-standing intellectual projects. To this end, he published *Determinism and Indeterminism in Modern Physics* (*Determinismus und Indeterminismus in der modernen Physik*) (1936), in which he defended a Neo-Kantian account of quantum mechanics. He also delivered two important lectures on culture during this period, including his inaugural address at the University of Göteborg, "The Concept of Philosophy as a Philosophical Problem" ("*Der Begriff der Philosophie als Problem der Philosophie*") (1935), and his presentation at the Warburg Institute in London, "Critical Idealism as Philosophy of Culture" (1936). Returning to his earliest projects in the history of ideas, Cassirer wrote the manuscript for a fourth volume of *The Problem of Knowledge*, dedicated to the developments in mathematics, physics, biology, and history after Hegel, which was published posthumously and translated in 1950 as *The Problem of Knowledge: Philosophy, Science, and History Since Hegel* (*Das Erkenntnisproblem in der Philosophie und Wissenschaft der neueren Zeit:Von Hegels Tod bis zur Gegenwart*). Meanwhile, still interested in Descartes, Cassirer wrote several pieces on Descartes in this period, which culminated in *Descartes: Teaching—Personality—Impact* (*Descartes: Lehre—Personlichkeit—Wirkung*) (1939).

His time in Sweden, however, also gave him the occasion to develop his thought in new moral and political directions. Immersing himself in the Swedish intellectual milieu, Cassirer wrote several pieces on prominent Swedish thinkers, like the philosopher and jurist, Axel Hägerström (1868–1939); the poet and feminist, Thomas Thorild (1759–1808); and Queen Christina (1626–1689).[9] Of these pieces, *Axel Hägerström* (1939) was particularly significant because in it, in addition to addressing Hägerström's theoretical, ethical, and legal thought, Cassirer expanded his *Philosophy of Symbolic Forms* in an explicitly practical direction, discussing morality and right (*Recht*).

In the spring of 1940 Cassirer also penned *The Logic of the Cultural Sciences* (*Zur Logik der Kulturwissenschaften*) (1942) in which he weighed in on the long-standing debate regarding the relationship between the 'natural sciences' (*Naturwissenschaften*) and the 'human sciences' (*Geisteswissenschaften*) or 'cultural sciences' (*Kulturwissenschaften*). He argued that both the natural sciences and human or cultural sciences count as sciences, and that the difference between them concerns the phenomena they study and conceptual frameworks they deploy. Cassirer also concluded with a final study, "The Tragedy of Culture," where he addresses an issue that would come to be of central importance for his late philosophy of culture, viz., the conflicts and tensions that give rise to the 'drama' of culture.

In 1941 Cassirer accepted a visiting position at Yale and while he planned to leave Sweden only temporarily, his steamer was the last the Germans permitted to sail between Sweden and the U.S. during

World War II. He would thus remain in the U.S. for the rest of his life. After several years at Yale, he spent 1944–1945 visiting at Columbia. And while walking in New York on April 13, 1944, Cassirer died of a heart attack at the age of 70.

In his years in the U.S. Cassirer taught both graduate and undergraduate courses;[10] served as a graduate advisor, notably for the philosopher of science, Arthur Pap (1921–1959); and was an active participant in the broader philosophical community in and around New Haven and New York. During this time Cassirer developed a close philosophical relationship with Susanne Langer (1895–1985), for whom Cassirer's work was highly influential.

Cassirer's last years were, as ever, highly productive. Although he had been encouraged by his American colleagues to undertake a translation of The Philosophy of Symbolic Forms, Cassirer chose instead to write a shorter introduction to his philosophy of culture in English, titled An Essay on Man (1944). Though in keeping with the spirit of The Philosophy of Symbolic Forms, Cassirer also presented An Essay on Man as a work in 'philosophical anthropology', oriented around the question, 'What is Man?'. The other major work in English that Cassirer wrote in this period was The Myth of the State, an early abbreviated version of which appeared in Fortune Magazine in 1944, and which was published posthumously in 1946. In The Myth of the State Cassirer offers a diagnosis of the rise of fascism in the 20th century in terms of a regression toward 'mythical thinking'. And although this is his best-known political work, it served as the capstone to his efforts extending back to Freedom and Form to develop an ethics and politics of culture.

Summary

Toward the end of his life, Cassirer remarked,

> Looking back on my long academic life I must regard it as a long Odyssey. It was a sort of pilgrimage that led me from one university to the other, from one country to the other, and, at the end, from one hemisphere to the other. This Odyssey was rich in experiences—in human and intellectual adventures.
>
> (quoted in Hendel 1949: 56)

Indeed, as we have seen throughout this survey of Cassirer's life and works, his journey from Berlin to Marburg, Hamburg to Oxford, Göteborg to New York, was as much an intellectual Odyssey as it was a physical one. In his early work, Cassirer's philosophy was animated by a Marburg concern with the nature of cognition, particularly as it manifested in mathematics and natural science. He pursued this project in a historically-oriented vein in the first two volumes of *The Problem of Cognition* and then in a more systematic vein in *Substance and Function*. Though Cassirer continued to devote attention to history of philosophy, philosophy of mathematics, and philosophy of science over the course of his career, in the 1920s he turned his energies toward developing a systematic philosophy of culture. While *The Philosophy of Symbolic Forms* was the crowning achievement of his early philosophy of culture, he further develops his philosophy of culture in the 1930s and 1940s, expanding it to include a philosophical anthropology and an ethics and politics of culture, which culminated in his final major works, *An Essay on Man* and *The Myth of the State*, respectively.

In the rest of this book, I explore the details of how Cassirer develops his philosophical system. I begin in the next chapter with a discussion of the Neo-Kantian framework within which he operates, before turning to his philosophy of mathematics and natural science in Chapters Three and Four, and his philosophy of culture in Chapters Five through Seven. In keeping with Cassirer's thought, this investigation will be wide-ranging; however, the guiding thread will be his fundamental commitment to thinking through the human being as someone who belongs to and builds her cultural world.

Notes

1 As will emerge, Cassirer eventually situates his account of mathematics and natural science in his philosophy of culture, so I return to these topics again in the chapters devoted to culture.

2 See Edgar (2020) for an overview of Cohen's philosophy.

3 In his memorial remarks, Fritz Saxl, Cassirer's close collaborator and friend, would note Cassirer's Olympian reputation and say, "Cassirer, Olympian and aloof, was yet the most humane and learned doctor of the soul. Higher praise could hardly be given to any man" (Saxl 1949: 51).

4 Throughout this study, I shall translate 'Erkenntnis' as 'cognition', reserving 'know-ledge' for 'Wissen', except in the case of the fourth volume of Das Erkenntnisproblem, which has been translated into English as The Problem of Knowledge.

5 See Chapter Two for a lengthier discussion of Cassirer's interpretation of the thing in itself.

6 Gawronsky reports that Cassirer's unifying idea for his philosophy of culture, viz., "symbolic form," had "flashed upon him" one day when he was stepping off a street car in Berlin in 1917 (Gawronsky 1949: 25).

7 See Bayer (2001), the essays in Cassirer Studies Volume III: The Originary Phenomena (2010), and Meland (2012/13) for a discussion of these manuscripts.

8 Cassirer and Heidegger had met as early as 1923 at a Kant-Gesellschaft in Hamburg and each engaged with the other's work prior to the debate. Heidegger cites PSFv2 in Being and Time, e.g., 51 fn. xi, and he wrote a review of PSFv2 in 1928. Cassirer cites Being and Time multiple times in PSFv3, e.g., 187 fn. 6/167–168 fn. 64/149 fn. 4, 220 fn. 3/184 fn. 86/163 fn. 2.

9 See, e.g., Cassirer's analysis of Thorild in "Thorild's Place in the Geistgeschichte of the 18th Century" ("Thorilds Stellung in der Geistesgeschichte des 18. Jahrhunderts") (1941) and "Thorild und Herder" (1941). For his treatment of Queen Christina, see Descartes: Lehre—Personalichkeit—Wirkung (1939) and the translation of three of Cassirer's 1938 lectures on Descartes and Queen Christina in Queen Christina and Descartes (Drottning Christina och Descartes) (1940).

10 Cassirer's courses included graduate seminars on philosophy of history, philosophy of science, and epistemology, and undergraduate courses on Kant, symbolism, aesthetics, Plato, and the theoretical foundations of culture (Symbol, Myth, and Culture Introduction 22).

Further reading

Cassirer, Toni (2003). Mein Leben mit Ernst Cassirer. Hamburg: Felix Meiner. A memoir written by Cassirer's wife, Toni Cassirer.

Eilenberger, Wolfram (2020). Time of the Magicians: Wittgenstein, Benjamin, Cassirer, Heidegger, and the Decade that Reinvented Philosophy. Translated by Shaun Whiteside. New York: Penguin. A portrait of Cassirer, alongside Ludwig Wittgenstein, Walter Benjamin, and Martin Heidegger, in the wake of World War I.

Ferrari, Massimo (2003). Ernst Cassirer. Stationen einer philosophischen Biographie. Translated by Marion Lauschke. Hamburg: Meiner. A discussion of Cassirer's overarching philosophical trajectory.

Friedman, Michael (2000). A Parting of the Ways: Carnap, Cassirer, and Heidegger. Chicago: Open Court. A discussion of the Davos disputation, which emphasizes its relevance for the analytic-continental divide in philosophy.

Gordon, Peter (2010). Continental Divide: Heidegger, Cassirer, Davos. Cambridge: Harvard University Press. A discussion of the Davos disputation, which emphasizes

Cassirer and Heidegger's fundamental disagreement over the nature of the human being as 'spontaneous' or 'thrown'.

Hansson, Jonas Hansson and Nordin, Svante (2006). *Ernst Cassirer: The Swedish Years*. Bern: Peter Lang. A discussion of Cassirer's time in Sweden.

Levine, Emily (2013). *Dreamland of Humanists: Warburg, Cassirer, Panofsky, and the Hamburg School*. Chicago: The University of Chicago Press. A discussion of Cassirer's time in Hamburg, his relationship with Warburg and Panofsky, and the history of the Warburg Library.

Schlipp, Paul Arthur, ed. (1949). *Library of Living Philosophers: The Philosophy of Ernst Cassirer*. Evanston: The Library of Living Philosophers. A collection of essays on Cassirer's philosophy by his contemporaries, which Cassirer was supposed to respond to, but was unable to do so before he died. This volume also includes a thorough biography of Cassirer, "Ernst Cassirer: His Life and His Work" by Dimitry Gawronsky and a set of memorial remarks about Cassirer after his death by Fritz Saxl, Charles Hendel, Edward Case, and Hajo Holborn.

Two
The Neo-Kantian framework

Introduction

For all the shifts and developments in Cassirer's body of work, his philosophical system remains, throughout, that of a Neo-Kantian. The term 'Neo-Kantian' is often used in a very broad sense to refer to philosophers who take the work of Immanuel Kant as their starting point. So understood, the school of Neo-Kantian philosophy is vast, including figures as diverse as Friedrich Schiller, Hermann von Helmholtz, Wilfrid Sellars, and Christine Korsgaard. There is, however, another narrower use of the term 'Neo-Kantian' that refers to a philosophical movement that dominated Germany from 1870 to 1920. A diverse range of thinkers participated in this development, including Helmholtz, Friedrich Albert Lange, Eduard Zeller, Otto Liebmann, and Alois Riehl. This movement eventually coalesced into two dominant schools: the so-called 'Marburg School' of Hermann Cohen, Paul Natorp, and Ernst Cassirer, and the so-called 'Southwest' or 'Baden School' of Wilhelm Windelband, Heinrich Rickert, and Emil Lask. The many differences between these thinkers notwithstanding, what united them all was a commitment to go 'Back to Kant!'.[1] In this chapter, my concern is with how Cassirer's affiliation with German Neo-Kantianism, in general, and Marburg Neo-Kantianism, in particular, laid the groundwork for his philosophy as a whole.

Before I proceed, though, a caveat about Cassirer's Kantianism is in order. While I focus here on the Kantian themes in Cassirer's philosophy, Kant is certainly not the only thinker who influences

Cassirer. Other figures, notably Leibniz, Goethe, Hegel, and Dilthey also play a decisive role in shaping Cassirer's philosophical outlook.[2] Indeed, part of what makes Cassirer's Neo-Kantianism distinctive is the way in which he refracts certain Kantian insights through the lenses provided by these other thinkers. In later chapters, I highlight the influence of Hegel and Dilthey in particular, but a fully comprehensive appreciation of Cassirer's philosophy requires attending to these other lines of influence as well.

In order to elucidate the Neo-Kantian foundation of Cassirer's philosophy, I begin in the next section by exploring Cassirer's general Kantian outlook. I open with a discussion of why the German Neo-Kantians, in general, thought a return to Kant was needed. As will emerge, these reasons were methodological in nature, as the Neo-Kantians sought to develop a philosophical method that could avoid the speculative metaphysics of German Idealism, on the one hand, and the empiricist method of positivism, on the other. And part of Cassirer's Kantian heritage turns on his embrace of these general methodological motivations. However, I also argue that Cassirer's return to Kant was precipitated by his embrace of Kant's so-called 'Copernican Revolution' in epistemology and ontology.[3] In this vein, I discuss Cassirer's embrace of a critical theory of cognition and critical idealism.

In the third section I take up Cassirer's relationship to Marburg Neo-Kantianism. Although Cassirer's earlier work in philosophy of mathematics and natural science is typically recognized as Marburg in character, there is a question of whether his later work in the philosophy of culture amounts to a break with his Marburg predecessors. However, there is a basic philosophical outlook that Cassirer inherits from Cohen and Natorp that guides him throughout his career. I describe this basic Marburg outlook in terms of two basic commitments. The first is a methodological commitment to the so-called 'transcendental method'. I show that Cohen and Natorp ultimately conceive of the transcendental method as a method for developing a critical philosophy of culture on an 'anti-psychologistic' basis. And I clarify the ways in which this method shapes Cassirer's body of work as a whole. The second commitment that Cassirer inherits pertains to a certain way of rereading Kant, which turns on an 'intellectualist' account of the relationship between the fundamental capacities of the mind, viz., 'sensibility' and 'understanding', and a 'regulative' reading of the thing in itself. This said, as I also discuss, Cassirer is critical of Cohen and Natorp for carrying forward the Marburg program in an overly rationalistic and restrictive way. Nevertheless, instead of breaking with the Marburg School of Neo-Kantianism on these grounds, Cassirer reenvisions it.

Over the course of this chapter, the general framework for Cassirer's philosophical system as a whole will thus emerge as one that involves a methodology, epistemology, ontology, and philosophy of mind that is indebted to Kant and the Neo-Kantians.

Back to Kant!

In this section, my focus is on the question, 'why Kant?' for Cassirer. Part of the answer requires looking at the motivations of the German Neo-Kantians, in general, and why they, Cassirer among them, thought philosophy needed to return to Kant. As we shall see, these motivations were tied to methodological considerations about the ongoing value of philosophy in the face of natural science. The other part of this answer turns on Cassirer's endorsement of Kant's so-called 'Copernican Revolution' in ontology and epistemology, which culminates in Cassirer's embrace of a 'critical idealism' and a 'critical theory of cognition'.

Neo-Kantians and the methodological crisis in philosophy

Though there are different accounts of why the Neo-Kantians, Cassirer included, felt the need to go 'back to Kant', in what follows I shall consider Cassirer's version of events. According to his reconstruction, the return to Kant was precipitated by a methodological crisis that arose in the 19th century concerning the on-going value of philosophy in light of progress in natural science.[4] Whereas prior to the 19th century, philosophy played an 'independent' role as a 'leader' in relation to natural science, by the mid-19th century it was no longer clear that this was the case (PK 10–11/12–13). While natural science was making rapid advances, it seemed to many that philosophy had become mired in the speculative metaphysics of German Idealists, like Fichte, Schelling, and Hegel. What is more, many viewed the *a priori* results of these idealist systems as out of touch with the empirical results of natural science. Making this point about the reception of Hegel, Cassirer says,

> In the domain of [natural] science . . . Hegel's system not only failed to disclose any [positive] result, but also led him and his disciples and successors to constant errors of judgment and pretensions [*ständigen Mißgriffen und Übergriffen*] that necessarily deprived speculative philosophy of any credit among empirical investigators.
>
> (PK 3/3, translation modified)

Given these doubts, so Cassirer's reconstruction continues, there developed a new trend in philosophy that eschewed *a priori* methodology altogether, in favor of an empiricist methodology, viz., positivism. According to the positivists, the only valid philosophical methodology is an empirical one in which philosophy takes its cue from the facts uncovered by the natural sciences. In this methodological framework, per Cassirer, philosophy occupies a position of "one-sided dependence" on science, in which "[i]nstead of assuming the role of leader . . . [philosophy] allow[s] itself to be led by the sciences and forced in a prescribed direction by each of them" (PK 15/17, 17/19). For the positivists, it was only by embracing this empiricist methodology that philosophy could hope to remain viable in relation to natural science.

For the Neo-Kantians, however, relegating philosophy to a subsidiary position in relation to natural science was no more attractive than doing speculative metaphysics. By their lights, the most desirable philosophical method is one that remains concretely grounded in the facts, on the one hand, and that acknowledges the independence of philosophy from natural science, on the other. And it is precisely this sort of method that they saw in Kant.

Kant, after all, insists that, "High towers and the metaphysically-great men that resemble them, around both of which there is usually much wind, are not for me. My place is the fertile *bathos* of experience" (Prol. 4:374). What is more, in his discussion of this "fertile *bathos*", Kant pays close attention to the 'facts' of natural science, specifically, of Newtonian mechanics. Nevertheless, Kant argues that if philosophy is to "enter upon the secure course of a science," it cannot do so by means of an empiricist methodology; it must embrace an *a priori* task, viz. giving an account of the *a priori* conditions of the possibility of experience, including the 'facts' of natural science (CRP Bvii). As Cassirer describes Kant's method,

> [Kant's] transcendental method has to assume the "fact of the sciences" [*Faktum der Wissenschaften*] as given, and seeks only to understand the possibility of this fact. . . . But even so, Kant does not stand merely in a position of dependence on the factual stuff of knowledge [*Wissenstoff*], the material offered by the various sciences. Kant's basic conviction and presupposition consists rather of this, that there is a universal and essential form of knowledge [*Wissens*], and that philosophy is called upon and qualified to discover this form and establish it with certainty.
>
> (PK 14/16–17)

For the Neo-Kantians, Kant's method thus promised philosophy a way out of its methodological crisis in the 19th century. In remaining oriented towards the 'facts', Kant's method could avoid the Scylla of speculative idealism. But by insisting on the *a priori* task of philosophy in relation to those facts, it could avoid the Charybdis of positivism.

Thus, according to Cassirer, the Neo-Kantian demand to go 'back to Kant' amounts to a demand to go back to Kant's method:

> The individual thinkers who belong to this movement [i.e., Neo-Kantianism] differ from each other in their interpretation of the Kantian doctrine as well as in the results which they reach from the Kantian premises. But, notwithstanding differences of detail, there is a certain methodological principle common to all of them. . . . [T]hey enquire into the possibility of *philosophy as a science*. . . . They take their cue from the *most general* statement of the Kantian problem . . ., to the fundamental aim of Kant, to lead philosophy "into the safe road of a science".
>
> ("Neo-Kantianism" 308, quoting CPR Bvii,
> emphasis in original)[5]

And part of the reason that Cassirer belongs to the Neo-Kantian movement is because he shares these general methodological concerns and commitments (more on Cassirer's method below).

Cassirer and the Copernican Revolution

In addition to these general methodological motivations, Cassirer was drawn to the basic epistemological and ontological picture that he takes Kant to defend as part of his so-called 'Copernican Revolution'. To clarify this line of influence, I first offer a sketch of the Copernican Revolution and the basic epistemological and ontological view that emerges from it, and then analyze how this Kantian paradigm shapes Cassirer's philosophical outlook as a whole.

Kant lays out the Copernican Revolution in the Preface to the second edition of the first *Critique*. He opens the Preface by raising a question I alluded to above, viz., whether philosophy can "enter upon the secure course of a science" (Bxiv–xv). Kant poses this as a question for metaphysics, in particular, which he understands as a discipline oriented toward *a priori* cognition, i.e., cognition 'prior to' experience. And his question is whether it is possible for these *a priori* endeavors in metaphysics to amount to a 'science'.

According to Kant, if we look at the traditional metaphysical offerings, e.g., rationalist theories of God, the soul, or the world, then we have little reason for hope. The telltale signs of being on a 'secure course of a science', which we find in other fields like mathematics and physics, seem to be missing in metaphysics. In metaphysics we find neither the unanimity, nor progress evident in these other 'secure sciences':

> In metaphysics we have to retrace our path countless times, because we find that it does not lead where we want to go, and it is so far from reaching unanimity in the assertions of its adherents that it is rather a battlefield . . .; on this battlefield no combatant has ever gained the least bit of ground, nor has any been able to base any lasting possession on his victory.
>
> (CPR Bxiv–xv)

Thus, in spite of however metaphysics might stylize itself, Kant argues that the current state of the field suggests it is far from entering the 'secure course of a science'.

However, before giving up hope, Kant claims that we might embrace a 'revolution' in philosophy, which he likens to Copernicus's revolution in astronomy:

> up to now it has been assumed that all our cognition must conform to the objects; but all attempts . . . on this presupposition, come to nothing. Hence let us try whether we do not get farther in the problems of metaphysics by assuming that the objects must conform to our cognition. . . . This would be just like the first thoughts of Copernicus, who, when he did not make good progress in the explanation of the celestial motions if he assumed that the entire celestial host revolves around the observer, tried to see if he might not have greater success if he made the observer revolve and left the stars at rest.
>
> (CPR Bxvi)

In this passage, Kant casts the revolution he calls for in philosophy as a 'Copernican' one because it involves rejecting traditional assumptions that have dominated the field. As Kant presents it, Copernicus made progress in astronomy possible by demanding that we reject our traditional assumptions about the relationship between the observer and the stars. Extending this revolutionary spirit to philosophy, Kant thinks that if philosophy is ever to progress towards being a science, we need to reject the traditional understanding of the relationship between cognition and objects.

Kant characterizes the traditional assumption that guides metaphysics as the idea that cognition conforms to objects, and he takes this assumption to carry with it both ontological and epistemological commitments. On the ontological front, Kant thinks that this assumption is underpinned by a view of the object of cognition as a mind-independent 'thing in itself' (*Ding an sich*) or 'noumenon'. And on the epistemological front, Kant maintains that this assumption is tethered to a view of cognition as a process in which we form a representation of a thing in itself that is given to us. However, according to Kant, on this picture of cognition, the relationship between the object that we cognize and our cognition is ultimately an 'empirical' relationship, in which we attempt to represent a mind-independent object that we are given (CPR A92/ B125). If this is right, then the possibility of achieving the sort of *a priori* cognition that traditional metaphysics is after is ruled out by the very ontological and epistemological picture that it adheres to. It is thus little wonder, Kant thinks, that the battle between different metaphysical camps amounts to a "mock combat," far from the path of secure science (B xv).

However, Kant claims that if we reject this traditional assumption and suppose, instead, that objects conform to cognition, then metaphysics may well find its way to the secure path of a science after all. Of course, it will take Kant the entire first *Critique* to fully vindicate this Copernican paradigm, but for my purposes, it will suffice to lay out the basic ontological and epistemological picture that Kant takes to be consonant with the Copernican Revolution, which, in turn, influences Cassirer.[6]

Kant labels his rival ontological picture 'critical' idealism.[7] According to critical idealism, the object of cognition is not a mind-independent 'thing in itself' or 'noumenon'; it is an 'appearance' (Erscheinung) or 'phenomenon', which conforms to the mind.[8] Kant explains this conformity in terms of appearances being 'conditioned' or 'determined' by the *a priori* 'forms' of the mind. His basic idea is that the formal structure of an appearance is something that is made possible by these *a priori* forms of the mind.

More specifically, in the first *Critique* Kant argues that appearances are conditioned or determined by the *a priori* forms of our two basic cognitive capacities, 'sensibility' (Sinnlichkeit) and 'understanding' (Verstand). According to Kant, these two capacities are characterized by certain *a priori* forms: sensibility by the pure intuitions of space and time, and the understanding by the twelve pure concepts, i.e., the 'categories', like 'substance' and 'causality'. So, for Kant, space, time, and the categories serve as the *a priori* forms of sensibility and understanding that condition or determine appearances. And his basic idea is that appearances have the formal structures that they do, for example, a spatio-temporal structure, a substance– accident structure, a causal structure, in virtue of the *a priori* forms of sensibility and understanding. At the heart of Kant's critical idealism is thus a view of the objects that we can cognize as appearances that depend on the *a priori* forms of sensibility and understanding.

In addition to this revolution in ontology, Kant defends a revolutionary epistemological theory of cognition, which I shall refer to as his 'critical theory of cognition'. Given his critical idealism, Kant's critical theory of cognition involves recognizing that cognition is a process in which we represent not absolutely mind-independent objects, but rather mind-dependent appearances. And this theory of cognition is one that he pursues at both an empirical and *a priori* level. On the empirical level, he offers an account of the empirical activities and representations involved in the 'empirical cognition', or 'experience' of appearances. Meanwhile, on the *a priori* level, he provides an analysis of how the *a priori* forms of sensibility and understanding condition or determine appearances. Through this *a priori* project, Kant claims that we can arrive at a body of *a priori* cognition that gives us cognition about appearances 'prior to' experience, but that does so without resorting to the traditional framework of metaphysics. It is thus to this body of *a priori* cognition, rather than traditional metaphysics, that Kant thinks we must turn in order to put philosophy on the 'secure path of a science'.

Though, Cassirer is by no means a Kantian to the letter, Kant's Copernican Revolution and the critical idealism and critical theory of cognition that accompany it lay the groundwork for Cassirer's philosophy as a whole. Indeed, Cassirer's commitment to this basic Kantian program unifies his early philosophy of mathematics and natural science and his later philosophy of culture. While the details of how exactly he works this out are a topic for later chapters, for now I want to highlight key moments in his early theoretical and later cultural writings that make clear the foundational role played by this Kantian framework.

Cassirer tends to frame his version of critical idealism and the critical theory of cognition in terms of a particular idiom, viz., that of 'substance' and 'function'. He labels an absolutely mind-independent thing in itself an 'absolute substance'. And he often refers to the forms that objects of cognition conform to as 'functions' (see, e.g., KLT 148–149/144–145). Hence the dialectic he sets up in *Substance and Function* between a substance-based thinking he rejects and function-based thinking he endorses (see Chapter Three). In this dialectic, the pre-Copernican position involves an ontological view of the object of cognition as an absolute substance, and an epistemological view according to which cognition is a process in which we form 'copies' of absolute substances. Cassirer calls this latter view the 'copy theory' (*Abbildtheorie*) of cognition (see, e.g., ETR 392/49, 433/97; SF 35/35, 164–165/178; PSFv1 3/3/74–75). Meanwhile, on the Kantian position, critical idealism amounts to the ontological view that the object of cognition is not an absolute substance, but rather an object that conforms to the functions of the mind. And the critical theory of cognition amounts to the epistemological view that empirical cognition involves the cognition of objects shaped by the functions of the mind, and *a priori* cognition involves cognition of these functions. In both his early theoretical philosophy and later philosophy of culture, Cassirer endorses a version of critical idealism and the critical theory of cognition.

Beginning with his early work in philosophy of mathematics and natural science, Cassirer outlines his commitment to critical idealism in *Substance and Function* as follows. He asserts that the object of cognition is not an "absolute substance beyond all cognition"; it is rather an object that has "*relative* being" (SF 297–298/321, emphasis in original, translation modified). More specifically, he asserts that an object of cognition is relative to the functions of 'progressing experience':

> If we determine the object not as an absolute substance beyond all cognition, but as the object shaped in progressing experience . . . it remains strictly within the sphere, which these principles [of logic] determine and limit, especially the universal principles of mathematical and scientific cognition. This simple thought alone constitutes the kernel of critical "idealism".
>
> (SF 297/321, translation modified)

In this context, Cassirer uses 'experience' in a technical way that is typical of Marburg Neo-Kantianism, viz., as a term to refer not to mundane everyday experience, but rather to the experience involved in mathematical-scientific cognition.[9] So when Cassirer claims that the object of cognition depends on functions of 'progressing experience', he means it depends on the *a priori* functions that make mathematical-scientific cognition possible. And it is these functions that he takes the object of mathematical-scientific cognition to be 'relative' to.

On the epistemological front, Cassirer's early historical and systematic writings are also animated by an effort to defend a critical theory of cognition against the 'copy theory' of cognition. In the first two volumes of The Problem of Cognition (*Das Erkenntnisproblem*) (1906, 1907), he argues that the historical development of modern European thought from the Renaissance to Kant can be understood in terms of a shift away from a copy theory of cognition and toward a critical theory of cognition. Meanwhile, taking up these themes from a systematic perspective in Substance and Function, which, recall, is subtitled "*Investigations of the Fundamental Questions of the Critique of Cognition,*" Cassirer argues that it is only a critical theory of cognition that can account for the possibility of cognition in mathematics and natural science. In this spirit, he then offers an account of the *a priori* functions that make cognition in these fields possible, which I shall take up in Chapters Three and Four.

If we turn to his philosophy of culture, we find that Cassirer's outlook continues to be shaped by a commitment to critical idealism and a critical theory of cognition, albeit in a more expansive form. Cassirer describes this expansion as follows at the outset of the first volume of The Philosophy of Symbolic Forms:

> When I attempted to apply my findings [from Substance and Function] . . . to the problems that concerned the *human sciences* [*Geisteswissenschaften*], it became increasingly clear to me that the general theory of cognition [*Erkenntnistheorie*] . . . would need to be fundamentally broadened. Rather than investigating only the general presuppositions of the scientific *cognizing* [*Erkennens*] of the world, it was equally necessary to differentiate the different basic forms of "*understanding*" [*Verstehen*] of the world and apprehend each one of them as sharply as possible in their distinctive tendency and spiritual [*geistigen*] form.
>
> (PSFv1 lxxix/vii/69, emphasis in original)

I want to highlight two features of how Cassirer reformulates his critical project in this passage. To begin, Cassirer makes clear that in the context of the philosophy of culture, the focus is not just on the 'facts' involved in mathematical-scientific cognition; he now orients himself towards the 'facts' uncovered through the 'human sciences' as well, for example, in history, art history, anthropology, sociology, and the like. On his view, these latter facts pertain to the various modes of 'understanding' at work in other domains of culture, like myth, religion, language, and art. To be sure, the cognition we achieve in mathematics and natural science continues to play a crucial role in his account of culture; however, on his view, in order to account for culture as a whole, one needs to investigate both the cognition achieved through mathematics and natural science and the understanding operative in the other domains of culture.

Furthermore, notice Cassirer's description of the 'forms' that underwrite culture as forms of 'spirit' (*Geist*). This is significant because in *The Philosophy of Symbolic Forms* Cassirer gives his Kantian account of the 'forms' or 'functions' of the mind a Hegelian-style twist, cashing them out as forms of 'spirit'.[10] In keeping with the general Neo-Kantian rejection of speculative metaphysics, Cassirer does not understand spirit as an absolute metaphysical substance. He, instead, conceives of 'spirit' as a kind of *a priori*, intersubjectively shared structure and activity, which is grounded in humanity. And in his philosophy of culture, he reformulates his critical idealism and critical epistemology with reference to the functions or forms of 'spirit'.

For example, he presents his critical idealism as follows in *The Philosophy of Symbolic Forms*:

> the Copernican revolution, with which Kant began, takes on a new and wider sense. It no longer refers only to the logical function of judgment but extends, with equal justification and right, to every tendency and every principle of spiritual configuration [*geistiger Gestaltung*]. . . . [T]he basic principle of critical thinking, the principle of the "primacy" of the function over the object, assumes in each special domain a new shape [*Gestalt*]. . . . With this, the critique of reason becomes a critique of culture.

It seeks to understand and demonstrate how the content [Inhalt] of culture . . . insofar as it is grounded in a general principle of form, presupposes an original act of spirit [ursprüngliche Tat des Geistes]. Herein the basic thesis of idealism finds its true and complete confirmation.

(PSFv1 8–9/8–9/79–80)

Cassirer here casts his philosophy of culture as an effort to widen the scope of the Copernican Revolution by transforming Kant's 'critique of reason' into a 'critique of culture'. To this end, he seeks to show how the basic critical idealist principle, the 'primacy of the function over the object', applies to all objects of culture, from myth and language to mathematics and natural science. And though this line of thought is familiar from his earlier work, he here frames the relevant functions as ones that belong to 'spirit'. The version of critical idealism that he thus defends in his philosophy of culture treats the objects of the cultural world as ones that conform to the functions or forms of the cognition and understanding involved in spirit.

Meanwhile, with respect to epistemology, as is evident in how he opens the first volume of The Philosophy of Symbolic Forms, in his philosophy of culture Cassirer expands his critical epistemology to encompass the 'understanding' involved in regions of culture, like myth, religion, and art., alongside the 'cognition' involved in mathematics and natural science. Yet even in this more expansive form, he remains wedded to the same basic epistemological commitments that animated his early theoretical work. He continues to reject a 'copy theory' of cognition and understanding in favor of a critical view:

Myth and art, language and science, are . . . not simple copies [Abbilder] of an available [vorhanden] reality but present [darstellen] the main directions of spiritual movement, the ideal process in which reality is constituted [konstituiert] for us as one and many.
(PSFv1 40–41/41/107, translation modified)

For Cassirer, the cognition and understanding involved in culture are not ways of 'copying' a mind-independent reality; they involve functions through which 'reality' is 'constituted' in the first place. And in keeping with Kant's epistemological aims, Cassirer endeavors to offer an account not only of how these different kinds of cognition and understanding take shape on an empirical level, but also of the *a priori* functions and forms of spirit that they involve.

Thus, even though in his philosophy of culture, Cassirer broadens his horizon beyond the facts of mathematical-scientific cognition to the facts of culture more broadly and presents this project as an investigation of the functions and forms of spirit, his project continues to be anchored in critical idealism and critical epistemology. And what this continuity in Cassirer's thought reveals is that he goes 'back to Kant' not just on account of Kant's methodology, but also because he sees in Kant's Copernican Revolution a critical ontology and epistemology that promises to account for the 'fertile *bathos*' of our mathematical, scientific, and cultural 'experience'.

Cassirer's Marburg Neo-Kantianism

Though Cassirer's appropriation of Kant accounts for key components of his Neo-Kantian framework, his is, ultimately, the framework of a Marburg Neo-Kantian. To be sure, as I discuss below, there are ways in which Cassirer distances his approach from Cohen and Natorp. Nevertheless, over the course his career, Cassirer shares a basic philosophical outlook with his Marburg predecessors, and, in this sense, he remains a Marburg Neo-Kantian throughout. I gloss this basic Marburg outlook in terms of two basic commitments. The first is a commitment to the so-called 'transcendental method', and the second is to a particular rereading of Kant. In what follows I examine how Cohen and Natorp frame this method and rereading of Kant, before turning to how Cassirer modifies and appropriates this outlook within his philosophy.

The transcendental method

As Natorp says in his article, "Kant and the Marburg School" (1912), the transcendental method is the "fixed point of departure" of Marburg Neo-Kantianism (Natorp 2015a: 181).[11] It can be formulated in terms of two steps: first, philosophy should orient itself around 'facts', such as, the 'facts of science', and, second, it should provide an account of the conditions that make those facts possible. In many ways, the transcendental method is a variation on the general methodological program I detailed in my discussion of Neo-Kantianism above. However, Cohen and Natorp have a particular conception of the relevant facts and conditions of possibility.

Beginning with their conception of the relevant facts, for Cohen and Natorp, these facts need to be understood in both historically dynamic and cultural terms. In the former vein, as Natorp puts it, the transcendental method constantly orients itself around "current, historically verifiable facts," which change over time (Natorp 2015a: 182). In insisting on the historically dynamic nature of these facts, they take their position to involve a revision of Kant's position. By their lights, Kant had a static understanding of the 'facts of science', as evidenced by him treating Euclidean geometry and Newtonian mechanics as the fixed facts of mathematics and natural science. On Cohen's and Natorp's view, by contrast, the facts change over time and philosophers thus have a responsibility to keep pace with this flux. As they often make this point, the facts are not simply given (geben), but given as a task (aufgegeben) (see Natorp 2015a: 183–184).

Cohen and Natorp, moreover, gloss the relevant facts as 'facts of culture'. This may come as something of a surprise because Marburg Neo-Kantianism is often characterized as a movement that is concerned exclusively with mathematics and the natural sciences. Indeed, Marburg Neo-Kantianism is often pitted against the Southwest School of Neo-Kantianism on this ground. According to this interpretation, Marburg Neo-Kantianism is a science-oriented school that is concerned with the 'natural sciences' (Naturwissenschaften), while the Southwest School is a value-oriented school that is concerned with the 'human sciences' (Geisteswissenschaften). And it is from this perspective that Cassirer's work on philosophy of culture seems to be a break not only with his predecessors, but also with the Marburg approach that characterizes his early theoretical work. Though prevalent, this way of interpreting both the Marburg and Southwest Schools is misleading.[12] While I cannot pursue the shortcomings of this characterization of the Southwest Neo-Kantians, like Windelband and Rickert,[13] with regard to the Marburg School, this scientistic characterization is in keeping neither with their reading of Kant, nor their own self-conception. Rather, according to Cohen and Natorp, the transcendental method, as Kant originally formulates it and as they execute it, is a method for elucidating the facts and conditions of culture.

Insisting on the Marburg commitment to the facts of culture, Natorp asserts,

> from the start, we [the Marburg Neo-Kantians] have viewed and described the philosophy of Kant—not to mention the philosophy of the transcendental method—as philosophy of culture, which we, embracing Kant's initiative, wish to work through in a more strict and consistent matter. We do not consider this philosophy of culture to be in opposition to philosophy of nature or natural science . . . [we] take natural science to be only one factor of human culture . . ., albeit an essential one. . . . [We see] Kantianism not just a matter for the head but also for the heart.
>
> (Natorp 2015a: 193–194)

For the Marburg Neo-Kantians, Natorp claims, the transcendental method takes as its starting point the facts of culture in general, that is, the "current, historically verifiable facts of science, morals, art, and religion" (182). Though the facts of mathematics and natural science are important for their philosophical program, they regard these as but one set of cultural facts among others. According to Natorp, then, the goal of the Marburg method is to clarify the conditions that make this sweep of cultural facts possible.

What is more, in directing their attention to culture, Cohen and Natorp do not take themselves to be departing from Kant, but rather carrying his program forward. On their interpretation, Kant's critical system should be understood as an attempt to elucidate various facts of culture: the first *Critique* is oriented toward the facts of mathematics and natural science, the second *Critique* toward the fact of the moral law, and the third *Critique* toward the fact of art.[14] By executing the transcendental method in relation to the facts of culture, Cohen and Natorp take their philosophical program to be in keeping with Kant's own.

In addition to this historical and cultural conception of the facts of culture, Cohen and Natorp endorse a particular conception of the 'conditions of possibility' of these facts, viz., an 'anti-psychologistic' conception of them. They thus take aim at a 'psychologistic' account of these conditions in terms of psychic acts, operations, or processes that occur in individual minds. And Cohen and Natorp insist that the 'conditions of possibility' that philosophy seeks should not be understood in these psychologistic terms.

On the Marburg view, the problem with psychologism is that it cannot account for an important feature of the 'facts' of culture, viz., the 'objective validity' of the judgments that we make in domains like mathematics, science, morality, and aesthetics.[15] For example, part of the 'fact' of physics is making judgments that are true of physical phenomena. However, as Natorp presents the objection, by adhering to psychologism, one "cancels out objective validity itself and changes it into purely subjective validity, if one attempts to support it on subjective grounds and to deduce it from subjective factors" (Natorp 1887 [2015b]: 168, translation modified). By restricting the conditions of the facts of culture to psychical acts, operations, or processes that take place in individual minds, Natorp claims that psychologism limits those conditions to what is 'merely subjective'. However, since something 'merely subjective' cannot ground something 'objective', Natorp objects that psychologism 'cancels out' the objective validity of the judgments, hence fails to do justice to the facts of culture it was meant to explain.

For this reason, Cohen and Natorp demand an anti-psychologistic analysis of the conditions of the facts of culture, which can accommodate the objective validity of the relevant judgments. By their lights, this requires giving an *a priori* account of these conditions, as having their source in 'pure' reason. Describing how this approach takes shape in Cohen's philosophy, Cassirer says,

> the critique of cognition [Erkenntniskritik] takes a strictly *objective* turn: it does not deal with representations and processes in the thinking individual, but with the validity relation [*Geltungszusammenhang*] between principles and "propositions" [*Sätzen*], which as such must be established independently of any consideration of the subjective psychological event of thinking.
>
> ("Hermann Cohen and the Renewal of Kantian Philosophy" 223/ECW 9, 123, translation modified)

Or, as Natorp puts it, the transcendental method has a "strict *objective* character," and it seeks to elucidate the "primordial law described quite aptly as the *logos, ratio,* and reason," which makes objectively valid judgments possible (Natorp 2015a: 182–183).

This being said, while Cohen and Natorp reject a psychologistic approach to the conditions of the possibility of culture, they do not dismiss the value of psychology altogether. Indeed, they pursue the possibility of developing a critical psychology, which provides a non-psychologistic account of the conditions of the possibility of the psychic processes, acts, and operations in individual minds. Though he was not able to complete it, Cohen intended to dedicate the fourth and final volume of his *System of Philosophy* to this topic. Meanwhile, the project of developing a critical approach to psychology is one that occupied Natorp throughout his career, for example, in *Introduction to Psychology According to Critical Method* (*Einleitung in die Psychologie nach kritischer Methode*) (1888) and *General Psychology According to the Critical Method* (*Allgemeine Psychologie nach kritischer Methode*) (1912). In rejecting psychologism, the Marburg Neo-Kantians do not, therefore, reject psychology.

In sum, on Cohen's and Natorp's view, the first step of the transcendental method involves philosophy orienting itself around the historically dynamic facts of culture, and the second step involves providing an anti-psychologistic account of the *a priori* conditions of possibility of these facts. So understood, their transcendental method is a method for developing a critical philosophy of culture.

If we now turn to Cassirer's methodology in this Marburg light, we find that throughout his career he remains committed to pursuing the transcendental method. In the first volume of *The Problem of Cognition*, Cassirer orients his historical analysis around the "historically developing" "'fact' of science" and seeks to elucidate the "logical function[s]" in which this fact originates (ECW 2, 14–15). Meanwhile in his early systematic writings, like "Kant and Modern Mathematics" and *Substance and Function*, he provides an account of the conditions of possibility of the facts of modern mathematics, including 19th century geometry and number theory, and of natural science, including classical mechanics and chemistry. In keeping with Marburg anti-psychologism, Cassirer claims that these conditions are ""transcendent" from the standpoint of the psychological individual," and must be understood as having an *a priori* source, viz., in the 'supreme principles' that are 'immanent' to 'logic' (SF 297/321).

As I mentioned above, although Cassirer's later philosophy of culture is read as a break with Marburg Neo-Kantianism, he sees it as methodologically continuous not just with Cohen's and Natorp's projects, but with Kant's. On the latter point, Cassirer argues that Kant's critical philosophy takes "the whole of spiritual culture [*geistigen Kultur*] as the *point of departure*" (KLT 155/150, translation modified, emphasis in original). And he regards this as a commitment that the Marburg School inherits. It should thus not surprise us that Cassirer continues to deploy the transcendental method in his philosophy on culture.

For example, endorsing the transcendental method in *The Philosophy of Symbolic Forms*, he maintains that he takes as his starting point "the empirically ascertained and secured facts [*Tatsachen*] of cultural consciousness," which are studied in the 'cultural sciences' (*Kulturwissenschaften*), like anthropology, history, linguistics, and then seeks to clarify the "conditions of . . . possibility" of these facts (PSFv2 13/13/11). As he puts it in 1936,

> [We] attempt to understand the facts . . . But that does not mean that [they] can be deduced in a mere *a priori* manner in thought. We have no other way to find [the facts] than to ask the special sciences, and we have to accept the data with which

we are provided by them, the data of the history of language, the history of art, and the history of religion. But what we are searching for are not the historical phenomena themselves. We try to analyze and to understand the fundamental modes of thinking, of conceiving, representing, imagining, and picturing that are contained in language, myth, art, religion, and even in science. . . . [W]e inquire into the nature of the different functions on which the phenomena, taken as a whole, depend.

> (CIPC 80–81/110, see also "What is 'Subjectivism'?"
> ("*Was ist 'Subjektivismus'*?") ECW 22, 169;
> AH 115–116; CP 56–57/148–151)

Moreover, in his philosophy of culture he continues to provide an anti-psychologistic account of the conditions of possibility of the facts of culture. To this end in the third volume of The Philosophy of Symbolic Forms, he contrasts the "empirical-causal explanation" of 'psychology' with his own critical approach to culture:

the philosophy of symbolic forms on the whole inquires not into the empirical source of consciousness. . . . Instead of pursuing its temporal causes of origin [Entstehungsursachen], the philosophy of symbolic forms is oriented solely toward what "lies within it"— toward the apprehension and description of its structural forms. Language, myth, and theoretical cognition are all taken as basic shapes [Gestalten] of "objective spirit," whose "being" it must be possible to disclose and understand purely as such.

> (PSFv3 56/54/49)

Here, Cassirer describes his method for approaching culture in anti-psychologistic terms, which looks not to the empirical mind of this or that individual, but rather to spirit as the 'pure' source of the conditions of the possibility of culture.

What is more, while Cassirer endorses the anti-psychologistic approach to culture of his Marburg predecessors, he also follows them in attempting to defend a critical psychology of culture. Critical psychology is, indeed, a topic Cassirer begins exploring already in the final chapter of Substance and Function ("On the Psychology of Relations"), and, as I discuss at length in Chapter Five, it plays a pivotal role in the third volume of The Philosophy of Symbolic Forms.

Stepping back, Cassirer's remark in the Davos disputation with Heidegger in 1929 is representative of his methodological approach as a whole:

> I stand by the Kantian posing of the question of the transcendental as Cohen repeatedly formulated it. He saw what is essential to the Transcendental Method in that this method began with a fact . . .: to begin with a fact in order to ask about the possibility of this fact.
>
> (cited in Heidegger 1990: 206, translation modified)

And insofar as Cassirer endorses the Marburg method throughout, his thought remains methodologically continuous with that of Cohen and Natorp.

This said, Cassirer nevertheless takes issue with the particular way in which Cohen and Natorp execute the transcendental method. In the first place, Cassirer objects that Cohen and Natorp tend to be overly restrictive in their understanding of the scope of the facts of culture. Cassirer suggests that Cohen and Natorp remain, in large part, wedded to Kant's tripartite focus on mathematical-natural science, morality, and aesthetics in the three Critiques, adding only religion to this list (see, e.g., PSFv3 64/62/56). By Cassirer's lights, a more expansive view of the facts of culture is required, one which takes into account facts like those pertaining to myth, language, technology, and history.

Moreover, Cassirer criticizes Cohen and Natorp for offering an overly rationalist account of the conditions of the possibility of the facts of culture. Cohen and Natorp analyze these conditions of possibility in terms of rational laws. As Natorp makes this point, the foundation of the facts of culture "remains the law . . ., which is ultimately that primordial law described quite aptly as the logos, ratio, and reason" (Natorp 2015a: 182). Although Cassirer agrees that laws serve as the condition of the possibility of some facts, for example, of mathematics and natural science, he does not think they condition all the facts of culture. By treating rational laws as the conditions of the possibility of all facts of culture, Cassirer objects that Cohen and Natorp fail to do justice to what is distinctive about other cultural regions, like myth, language, and art. As an alternative, Cassirer advocates for the more encompassing notions of 'function' and 'form' as the key to unlocking the conditions of possibility of culture as a whole, reserving the notion of rational law for regions like mathematics, natural science, and right (see PSFv3 64–65/62–63/56–57).

In these two ways, Cassirer distances his execution of the transcendental method from Cohen and Natorp's. However, this is a distance operative within a shared methodological framework, and, in this sense, Cassirer's methodology remains that of a Marburg Neo-Kantian.

Rereading Kant

The second aspect of the basic Marburg outlook that Cassirer inherits from Cohen and Natorp pertains to their rereading of Kant. On this issue, it is worth pointing out that for the Marburg Neo-Kantians, being a Kantian does not require defending Kant to the letter. It is a matter of embracing his methodology and revising 'the letter' when need be. As Natorp makes this point:

> According to the classical meaning of the term, philosophy is the eternal striving for fundamental truth, not the aspiring to possess it. Kant in particular, grasping philosophy as critique and as method, wanted to teach philosophizing rather than "a" philosophy. He who thinks otherwise is a poor student of Kant!
> (Natorp 2015a: 180)

Their appropriation of Kant thus involves a rereading of certain key Kantian claims, specifically Kant's account of the mind and his theory of the thing in itself.

With respect to Kant's philosophy of mind, Cohen and Natorp reject a dualistic reading of Kant's account of sensibility and understanding as the "two stems" or "two fundamental sources" of cognition (CPR A15/B29, A50/B74).[16] Given Kant's dualistic characterization of these two faculties (as passive and active, receptive and spontaneous, intuitive and conceptual, sensing and thinking), it can be tempting to read them as two distinct capacities that are independent from one another and that collaborate together in cognition.

However, Cohen and Natorp take issue with this sort of dualistic picture of the human mind. Describing Cohen's position on this issue, Cassirer says,

> Cohen rejected a fundamental step in Kant's theory of know-ledge, for he did not admit the distinction between "sensibility" and "understanding" which is a cardinal point in Kant's *Critique of Pure Reason*. Sensibility is described by Kant as the "receptivity" of the human mind, whereas understanding is defined as a form of spontaneity. But according to Cohen we have to efface the term "receptivity" from our theory of knowledge. Neither in its sensuous experience nor in its rational activity is the mind a *tabula rasa*, an empty tablet upon which outward things make their impressions. It is active in all its functions, in perception as well as conception, in feeling as well as in volition.
>
> ("Hermann Cohen" 226/ECW 24, 167)

As Cassirer indicates in this passage, Cohen attributes to Kant a dualistic conception of sensibility and understanding, and Cohen argues that this is a mistake: there is no purely receptive capacity in the mind. On Cohen's view, the activities and representations ('intuitions') of sensibility have their source in the understanding. In a similar vein, Natorp claims, for the Marburg School, "'intuition' no longer remains a cognitive factor that stands across from or is opposed to thinking. It *is* thinking" (Natorp 2015a: 186). Cohen and Natorp thus defend, what Cassirer calls, a 'logicist', or, what has more recently been called, an 'intellectualist' or 'conceptualist' account of sensibility, according to which its activities and representations have their source in the understanding, qua the spontaneous capacity for conceptual thought (see LCS 65/423). I shall refer to this as Marburg 'intellectualism'.

The Marburg version of intellectualism is one that, in turn, shapes their analysis of sensibility at both the *a priori* and empirical level. At the *a priori* level, Cohen and Natorp argue that the 'pure intuitions' of space and time have their source in the understanding and, as such, are to be understood as forms or functions of thought. Hence Natorp's claim that, "in the basic determinations of time and space, thinking takes shape in a downright typical fashion *as a 'function*,' rather than as an 'intuition'" (Natorp 2015a: 185).[17]

Meanwhile, at the empirical level, Cohen and Natorp argue that even our most basic sensory encounters with the world, for example, in empirical intuition and perception, involve activities and representations that have their source in the spontaneity of the understanding. As we saw Cassirer render Cohen's position above, "in its sensuous experience . . . the mind [is not] a *tabula rasa*, an empty tablet upon which outward things make their impressions. It is active in . . . perception." In endorsing this active account of empirical intuition and perception, Cohen and Natorp call for a revision of Kant's account of 'sensation'.[18] According to Kant, our sensory encounters are mediated by 'sensation', which is, "The effect of an object on the capacity for representation, insofar as we are affected by it" (CPR A19–20/B34). This passively received sensation serves as the given 'matter' of cognition, which is then actively shaped through the 'forms' of cognition. However, according to Cohen and Natorp, there is no wholly receptive moment of sensibility, no 'unformed matter' that is brutely given. Even the most basic aspects of sensory consciousness, they claim, are grounded on the conceptual forms of thought. Cohen and Natorp thus defend an intellectualist account of our empirical encounters with the sensible world as dependent on conceptual thought from the ground up.

When we look at Cassirer's philosophy of mind, we find that he too rejects a dualistic account of sensibility and understanding in favor of a kind of intellectualist picture, according to which sensibility is, in fact, a manifestation of a spontaneous capacity of the mind. His reading of Kant on this issue is subtle. According to Cassirer, at the outset of the first *Critique*, for example, in the Transcendental Aesthetic, Kant does present sensibility and understanding in dualistic terms; however, Cassirer argues that over the course of the Transcendental Analytic Kant revises his view. Cassirer maintains that in the Transcendental Deduction, Kant acknowledges that the "unity between [sensibility and understanding] is . . . grounded . . . in an originary-function of theoretical knowing [*Urfunktion des theoretischen Wissens*]," viz., "the synthetic unity of apperception" (PSFv3 9/9/8). Cassirer also points to Kant's remarks in the Amphiboly to the effect that 'matter' and 'form' are "concepts of reflection," that we form

by abstracting the idea of matter and form from their more primary unity (PSFv3 11/11/10). From Cassirer's point of view, Kant's
seemingly dualistic claims in the Aesthetic are ones Kant, in fact,
revises in the Analytic.

Moreover, like his Marburg predecessors, Cassirer endorses a
kind of intellectualist account of the unity between sensibility and
understanding. At the *a priori* level, in his early work he treats space
and time as forms of thought, describing them as 'concepts', 'categories', 'rules of the understanding', and 'logical invariants' (see,
e.g., SF Ch. 4.6, 269/290, ETR 420/81). Later in *The Philosophy of
Symbolic Forms* he again calls space and time 'concepts', though he now
traces these concepts to spirit (PSFv3 15/14–15/13). In this vein,
he describes space and time as "basic relations [Grundrelationen]" of
spirit, which make "spiritual combination [geistigen Zusammenfassung]"
possible (PSFv1 25/25/94, 28/29/97). Though he now identifies
spirit rather than thought as the spontaneous source of space and
time, he remains committed to the idea that they have their source
in something spontaneous, rather than receptive.

Cassirer also defends a version of intellectualism in his account of
sensibility and understanding in his empirical account of intuition
and perception. Again, in keeping with his Marburg predecessors, he
rejects the idea of a sensation as the 'unformed matter' of intuition
or perception:

> the possibility disappears of separating the "matter" of cogni-
> tion from its "form" by referring them each to a different origin
> in absolute being. . . . Matter *is* only with reference to form,
> while form, on the other hand, *is valid* only in relation to matter.
> (SF 310–311/335–336, translation modified)

For Cassirer the 'matter' of empirical intuition and perception
necessarily depends on 'form'. Taking up this theme in *The Philosophy
of Symbolic Forms*, he argues that it is a mistake to posit any purely
material layer of sensory consciousness that lies outside of all form:

> We never find "naked" sensation, as *materia nuda*, to which some
> form-bestowing [Formgebung] is adjoined—rather, all that is
> graspable and accessible to us is the concrete determinacy, the

living multiformity, of a perceived world, which is dominated and permeated through and through by certain modes of forming [Formung].

(PSFv3 16–17/17/15).

According to Cassirer, there is thus a moment of spontaneity involved in empirical intuition and perception from the outset.

This said, Cassirer distances his intellectualism from a judgment-centric version of it, according to which intuition and perception require judgment. Although we can make judgments about what we intuit or perceive, Cassirer argues that empirical intuition and perception involve a kind of spontaneity that does not require judgment. Rather, on his view, our sensible capacities themselves are spontaneous. That is to say, the activities and structures involved in intuition and perception are expressions of our spontaneity. As such, whenever we exercise those capacities, we will be engaging in a spontaneous act. Quoting Goethe to this end, Cassirer claims that in intuition and perception we are already "'seeing with the eyes of the spirit'" (PSFv3 155/150/134).[19]

In these ways, Cassirer takes over the Marburg intellectualist rereading of sensibility, according to which the allegedly receptive aspects of sensibility have their seat in our spontaneity, at both the *a priori* and empirical levels. And he, thereby, rejects a dualistic account of the mind in the Marburg style.

The second element of the Marburg rereading of Kant that is significant for Cassirer is their interpretation of the 'thing in itself' (Ding an sich). Though there is a great deal of debate about how to read Kant's account of the thing in itself, he, at least at times, appears to treat it in metaphysical terms, as something that grounds appearances and that affects us.[20] However, Cohen and Natorp reject a metaphysical reading of the thing in itself. On their view, we cannot meaningfully talk about some brutely given thing in itself that affects us; we can only talk about objects that are dependent on cognition. They, instead, endorse a 'regulative' reading of the thing in itself as an idea that 'regulates' our empirical inquiry. To this end, they draw on Kant's account of regulative principles as principles that heuristically guide empirical inquiry toward the goals of systematicity and completeness (see CPR A509/B537, A671/B699). According to Cohen and Natorp, the thing in itself should be understood along these lines, as an "endless task [*ewige Aufgabe*]" that we strive towards, viz., the task of developing a systematic and complete "sum total [*Inbegriff*] of scientific cognition" (Natorp 1918: 19, my translation).

Cassirer also follows his Marburg teachers in this vein, rejecting a metaphysical reading of the thing in itself in favor of a regulative one. Indeed, in *The Philosophy of Symbolic Forms*, Cassirer claims that metaphysical questions about the thing in itself are not ones we can meaningfully answer, indeed, even ask:

> the question . . . as to what absolute actuality is beyond the totality [*Gesamtheit*] of these spiritual functions, as to what the "thing in itself" might in this sense be, this question, of course, cannot be answered except to say that we learn more and more to recognize it as an ill-formed problem, as a hallucination [*Trugbild*] of thinking.
>
> (PSFv1 45/46/111)

Or, as he puts it in *Substance and Function*, the metaphysical question of the thing in itself involves a 'fallacy', a "*petitio principia*" (SF 310/335). Cassirer thinks the metaphysical question of the thing in itself is ill-formed because it involves asking whether the thing in itself stands in a causal relation to the phenomenal world, a question that makes no sense given Kant's argument that causality has significance only as a category that governs the phenomenal world. Embracing the Marburg alternative, on Cassirer's view, the only value the thing in itself has is as a regulative demand, viz., as the demand,

> to apprehend in one *system* the diverse methodological directions of knowledge [*Wissens*] in all their recognized particular natures and autonomy, such that the individual members, precisely through their necessary diversity, will complement and require one another.
>
> (PSFv1 5/5–6/77, emphasis in original)

As was the case with the Marburg rereading of sensibility and understanding, Cassirer thus follows Cohen and Natorp in endorsing a regulative reading of the thing in itself. And this way of rereading Kant, which sets the parameters of Cassirer's philosophy of mind and his philosophy of science, is another significant point of continuity between his project and that of his Marburg predecessors.

Summary

In this chapter, I explored the basic Neo-Kantian framework that serves as the foundation of Cassirer's philosophical system. As I noted at the outset, although other thinkers, like Leibniz, Goethe, Hegel, and Dilthey, influence Cassirer's outlook, a certain set of Kantian and Neo-Kantian commitments lay the groundwork for his philosophy as a whole. In order to clarify these commitments, I began with a discussion of the general methodological reasons the Neo-Kantians were drawn to Kant, as promising a middle path between speculative idealism and positivism. Though Cassirer is moved by these considerations, I also addressed why he is drawn to Kant in light of the Copernican Revolution. To this end, I examined the impact that Kant's critical idealism and critical epistemology had on Cassirer's early philosophy of mathematics and natural science and his later philosophy of culture.

In addition to these Kantian commitments, I explored the influence of Marburg Neo-Kantianism on Cassirer. As I discussed, although Cassirer is critical of Cohen and Natorp for being overly restrictive and rationalistic in focus, this does not amount to a rejection of Marburg Neo-Kantianism altogether. Instead, I showed that Cassirer's basic outlook remains Marburg in nature. To bring out his on-going Marburg commitments, I considered Cassirer's appropriation of the transcendental method, as a method that orients philosophy around the historically dynamic facts of culture and that provides an anti-psychologistic account of the a priori conditions of possibility of those facts. I also analyzed the impact of the Marburg rereading of Kant on Cassirer. In this vein, I highlighted Cassirer's intellectualism and his rejection of a dualistic account of sensibility and understanding in his philosophy of mind and his regulative interpretation of the thing in itself.

On the basis of this Neo-Kantian framework, Cassirer develops a philosophical system intended not only to keep pace with the revolutionary developments in mathematics and physics in the 19th and 20th centuries, but also to do justice to who we are as beings who create a shared cultural world. And it is to the details of this system that we shall now turn.

Notes

1 This phrase was popularized by Otto Liebmann in *Kant and the Epigoni* (1865), in which he criticizes post-Kantian idealism and concludes each chapter with the phrase, *"Also muss auf Kant zurückgegangen warden."* As Willey (1978): 80 and Köhnke (1991): 128 note, this phrase does not originate with Liebmann, but had been used as early as 1845. Liebmann's efforts were also indebted to his Kantian teacher, Kuno Fischer, as well as Eduard Zeller's 1862 speech urging a return to Kant (published as *Über Bedeutung und Aufgabe der Erkenntnisstheorie* (1862)).

2 For Leibniz's influence on Cassirer, see Ferrari (2002); Rudolph (2003), (2008); Seidengart (2012); and Moynahan (2013). For Hegel's influence, see Kreis (2010) and Verene (2011). For Dilthey's influence, see Rudolph (2003) and the other essays in Leinkauf, ed. (2003) and Kreis (2010). For Goethe's influence, see Naumann and Recki (2002).

3 As noted in Chapter One, I translate 'Erkenntnis' as 'cognition' and 'Wissen' as 'knowledge', except in the case of the title of the fourth volume of *Das Erkenntnisproblem*, which was published in English as *The Problem of Knowledge*.

4 See, e.g., PK: Introduction; "Neo-Kantianism" in the *Encyclopedia Britannica* (1929): 308; "Hermann Cohen and the Renewal of Kantian Philosophy" (1912): 221–222/ECN 9, 122–123.

5 References to the *Critique of Pure Reason* (CPR) are to the A and B pagination of the first (1781) and second (1787) editions (A/B). All other references to Kant are to the volume and page number of *Kants gesammelte Schriften*.

6 In what follows, I assume that in virtue of endorsing a kind of idealism, Kant commits himself to a certain ontological view concerning the objects of cognition.

7 Kant also refers to his position as 'transcendental' or 'formal' idealism, but Cassirer tends to follow Kant's remark at Prol. 4:292, to the effect that the label of 'critical' idealism is better than 'transcendental' idealism, and I follow suit.

8 I return to the question of the thing in itself below in my discussion of the Marburg reading of it.

9 For the Marburg use of 'experience' in this sense, see Richardson (2003) and Holzhey (2005).

10 See Chapter Five for a discussion of the influence of Hegel and Dilthey on Cassirer's conception of spirit.

11 For Cassirer's discussion of the role the transcendental method plays in Kant, see e.g., PK 14–15/16–17, CP 52–55/144–148, PSFv3 55–56/53–54/49. For Cassirer's emphasis on the role the transcendental method plays in Cohen's

and Natorp's philosophies, see "Hermann Cohen and the Renewal of Kantian Philosophy" (1912), "Neo-Kantianism" (1929): 311–312, "Paul Natorp" (1925), and "Hermann Cohen" (1943).

12 For discussion of the relationship between the two schools, see Krijnen and Noras (2012), Heis (2018), Clarke (2019).

13 See, e.g., Windelband's "History and Natural Science" (Geschichte und Naturwissenschaft) (1915 [1980]) and Rickert's Cultural Science and Natural Science (Kulturwissenschaft und Naturwissenschaft) (1926, 6th/7th ed.) and The Limits of Concept Formation in Natural Science (Die Grenzen der naturwissenschaftlichen Begriffsbildung) (1929 [1986], 6th ed.).

14 This interpretation is the one Cohen develops at length in his three volume Kant-interpretation: Kant's Theory of Experience (Kants Theorie der Erfahrung) (1871/1885), Kant's Grounding of Ethics (Kants Begründung der Ethik) (1877), and Kant's Grounding of Aesthetics (Kants Begründung der Aesthetik) (1889).

15 See Anderson (2005) for a discussion of anti-psychologism in Neo-Kantianism in general, and Edgar (2008) for a discussion of Natorp's anti-psychologism in particular.

16 There is much on-going debate about whether to attribute to Kant a dualistic view of sensibility and understanding. See McLear (2020) for an overview.

17 As the Marburg Neo-Kantians tend to emphasize, although in the Transcendental Aesthetic Kant indicates that space and time belong to sensibility, his later comments about space and time in the B Deduction (B160–161fn.) might suggest that he conceives of them as the products of the understanding.

18 See Baumann (2019) for discussion of Cohen's view of sensations.

19 In order to emphasize the pre-discursive nature of this kind of perceiving, Cassirer highlights the Kantian idea that perception depends on imagination (see PSFv3 154–155/149–150/134). See Matherne (2014) and Kreis (2015) for a discussion of the extent of Cassirer's intellectualism about perception. See Endres (2020) for a discussion of Cassirer's phenomenology of perception as a whole.

20 For Cassirer's discussion of Kant's position on the thing in itself, see KLT 213–217/206–209, 255–263/245–253, 359–360/345–346) and Das Erkenntnisproblem Bd. 2 ECW 3, 733–762/613–638. On his reading of Kant, although Kant's considered view is that the thing in itself is a 'norm', 'goal', or 'task', Kant presents his view in a "paradoxical and ambiguous" way that invites a metaphysical interpretation (KLT 256/247).

Further Reading

Beiser, Fred (2014). The Genesis of Neo-Kantianism, 1796–1880. Oxford: Oxford University Press. A discussion of the early developments in Neo-Kantianism prior to the Marburg School.

Ferrari, Massimo (2009). "Is Cassirer a Neo-Kantian Methodologically Speaking?" In *Neo-Kantianism in Contemporary Philosophy*, edited by Rudolf A. Makkreel and Sebastian Luft. Bloomington: Indiana University Press: 293–314. A discussion of Cassirer's methodological commitment to Marburg Neo-Kantianism.

Keller, Pierre (2015). "Cassirer's Retrieval of Kant's Copernican Revolution in Semiotics." In *The Philosophy of Ernst Cassirer: A Novel Assessment*, edited by J. Tyler Friedman and Sebastian Luft, 259–288. A discussion of the role signs and other semiotic themes play in Cassirer's appropriation of Kant's Copernican Revolution.

Köhnke, Klaus (1991). *The Rise of Neo-Kantianism: German Academic Philosophy Between Idealism and Positivism*. New York: Cambridge University Press. A discussion of the history of Neo-Kantianism.

Luft, Sebastian (2015). *The Space of Culture: Towards a Neo-Kantian Philosophy of Culture*. Oxford: Oxford University Press. A discussion of the Marburg School's commitment to the philosophy of culture.

——ed. (2015). *The Neo-Kantian Reader*. London: Routledge. A reader that includes primary texts by a wide range of Neo-Kantians.

Matherne, Samantha (2015). "Marburg Neo-Kantianism as Philosophy of Culture." In *The Philosophy of Ernst Cassirer: A Novel Assessment*, edited by J. Tyler Friedman and Sebastian Luft, 201–231. Berlin: De Gruyter. A discussion of the Marburg School's commitment to the philosophy of culture.

——(2018). "Cassirer's Psychology of Relations: From the Psychology of Mathematics and Natural Science to the Psychology of Culture." *Journal of the History of Analytic Philosophy* Special Issue: Method, Science, and Mathematics: Neo-Kantianism and Early Analytic Philosophy, edited by Lydia Patton and Scott Edgar, 6(3): 132–162. A discussion of the psychology Cassirer develops in *Substance and Function* and *The Philosophy of Symbolic Forms*.

Moynahan, Gregory (2013). *Ernst Cassirer and the Critical Science of Germany, 1899–1919*. London: Anthem Press. A discussion of the development of Cassirer's early philosophy as it develops in relation to Cohen and Leibniz.

Renz, Ursula (2002). *Die Rationalität der Kultur: zur Kulturphilosophie und ihrer transzendentalen Begründung bei Cohen, Natorp und Cassirer*. Hamburg: Felix Meiner. A discussion of the Marburg School's commitment to the philosophy of culture.

——(2005). "Critical Idealism and the Concept of Culture: Philosophy of Culture in Hermann Cohen and Ernst Cassirer." In *Hermann Cohen's Critical Idealism*, edited by Reinier Munk, 327–356. Dordrecht: Springer. A discussion of Cohen's and Cassirer's commitments to the philosophy of culture.

Richardson, Alan (2006). ""The Fact of Science' and Critique of Knowledge: Exact Science as Problem and Resource in Marburg Neo-Kantianism." In *The Kantian Legacy in Nineteenth-century Science*, edited by Michael Friedman and Alfred Nordmann, 221–226. Cambridge, MA: MIT Press. A discussion of the transcendental method in the context of Marburg philosophy of science.

Rudolph, Enno (2003). *Ernst Cassirer im Kontext*. Tübingen: Mohr Siebeck. A discussion of the influences on Cassirer's philosophy beyond Kant, including Plato, Renaissance philosophy, Leibniz, Herder, Schleiermacher, and Dilthey.

——(2008). "Symbol and History: Ernst Cassirer's Critique of the Philosophy of History" in *The Symbolic Construction of Reality: The Legacy of Ernst Cassirer*, edited by Jeffrey Andrew Barash. 3–16. A discussion of how Cassirer's philosophy departs from Kant's, with emphasis on Leibniz's influence on Cassirer.

Three
Philosophy of mathematics

Introduction

Many of Cassirer's most significant philosophical contributions emerge as the result of his efforts to defend a Neo-Kantian account of modern mathematics and (mathematical) natural science.[1] In keeping with his Neo-Kantian outlook, Cassirer pursues this project by applying the transcendental method to the historically developing facts of modern geometry, number theory, physics, and chemistry. For a philosopher with Kantian leanings, these facts pose a particular challenge. To many, it seems that the Kantian project is so wedded to the exclusive truth of Euclidean geometry and Newtonian mechanics that the revolutionary developments in non-Euclidean geometry, relativity, and quantum mechanics invalidate a Kantian approach to mathematics and natural science altogether. Sensitive to this skepticism, Cassirer seeks to revise and update the Kantian approach in order to make sense of these radical paradigm shifts. This goal orients his major works, *Substance and Function* (1910), *Einstein's Theory of Relativity* (1921), the third volume of *The Philosophy of Symbolic Forms: Phenomenology of Cognition* (1929), and *Determinism and Indeterminism in Modern Physics* (1936).

My aim in the next two chapters is to explore how Cassirer develops an account of the revolutionary advances in mathematics and natural science within his Neo-Kantian framework. In the next part of this chapter, I discuss the particular idiom in which Cassirer presents this framework, viz., the idiom of 'substance' and 'function'. I then dedicate the rest of this chapter to his philosophy of mathematics and the next to his philosophy of natural science. However, given that he ultimately regards mathematics and natural science as part of the broader cultural world, I shall return to these topics again in my discussion of his philosophy of culture in Chapters Five, Six, and Seven.

In this chapter, after addressing Cassirer's 'substance-function' framework, I devote my attention to his philosophy of mathematics. In keeping with the transcendental method, he takes as his point of departure the "new 'fact'" of "modern mathematics," a fact that includes advances in geometry, like non-Euclidean geometry, projective geometry, and Klein's Erlangen Program, and advances in number theory, like theories of irrational, imaginary, and transfinite numbers (KMM 65/31).[2] And he sets himself the task of explicating the conditions that make these facts possible. As we shall see in what follows, in accounting for these conditions, Cassirer develops a version of the critical theory of cognition and critical idealism, which culminates in a position I refer to as 'logical structuralism'. According to Cassirer's logical structuralism, mathematics is the science of ideal structures that are constructed on the basis of functions or relations, which have their source in logic. Although in defending logical structuralism, Cassirer distances his view of mathematics from intuition-based views, like Kant's, he also distinguishes his view from the logicism of Frege and Russell. Though, like Frege and Russell, Cassirer denies that intuition has a role to play in the foundation of mathematics, unlike Frege and Russell, he nevertheless takes intuition to make an essential contribution to mathematics vis-à-vis the role intuition plays in the application of mathematics to the physical world in natural science. By thus balancing logical structuralism with an account of the applicability of mathematics, Cassirer's philosophy of mathematics aims to acknowledge both the logical foundation of mathematics and the instrumental role it plays in the scientific cognition of nature.

In order to explore Cassirer's philosophy of mathematics, I begin with an overview of the specific function-based view of mathematics that he defends, viz., logical structuralism. For Cassirer, logical structuralism is both a foundational theory, according to which the foundation of mathematics is logical in nature, and an ontological theory, according to which mathematical objects are ideal structures that are generated on the basis of functions or relations.

I then consider his argument that the 'new' facts of 'arithmetic'[3] and geometry give us reason to endorse logical structuralism. The work of Richard Dedekind plays a prominent role in this argument, as Cassirer sees in Dedekind a defense of logical structuralism, which serves as the model for how we should understand not only the 'new' facts of mathematics, but also the foundations and objects of arithmetic and geometry more generally. In addition to Dedekind, I consider the central role that the invariant-based method of Jean-Victor Poncelet and Felix Klein and the axiomatic method of David Hilbert play in Cassirer's case for why we should embrace logical structuralism in geometry.

I close the chapter by addressing Cassirer's account of the relationship between intuition and mathematics, especially in light of the so-called 'foundational crisis' in mathematics in the early 20th century. I examine his criticisms of the logicism of Frege and Russell in this context, and the different roles that Cassirer thinks intuition can play in mathematics. First, however, we need to turn to Cassirer's 'substance-function' framework.

Substance and function

Cassirer's approach to philosophy of mathematics and natural science, in general, hinges on his embrace of Kant's revolutionary 'Copernican' idea that cognition in these fields does not conform to objects, but rather objects conform to it (CPR Bxvi). As I noted in Chapter Two, Cassirer articulates the Copernican Revolution in terms of a shift from 'substance'- to 'function'-based thinking. He thus characterizes the pre-Copernican approach as one that involves an ontological account of the objects of cognition as 'absolute substances' that are wholly mind-independent, and an epistemological account of cognition as a process in which our minds 'copy' those absolute substances. By contrast, Cassirer casts the Kantian position in terms of the recognition that objects of cognition conform to the 'functions' of the mind. Hence Cassirer's gloss of "the basic principle of critical thinking" as "the principle of the 'primacy' of the function over the object" (PSFv1 8/9/79). And this is why

Cassirer presents his rival ontological theory, critical idealism, as a theory about objects of cognition conforming to functions of the mind and why he orients his epistemological theory, the critical theory of cognition, around an analysis of these functions.

Although Cassirer draws on the substance-function framework in his later philosophy of culture, his canonical analysis of it is to be found in *Substance and Function*. The original German title of this book is *Substanzbegriff und Funktionsbegriff*, which literally translates to 'Substance-concept and Function-concept', and Cassirer chooses this title because *Substance and Function* is, in large part, a book about concepts. Cassirer takes this topic as his guiding thread because he thinks that debates about how we achieve cognition in mathematics and natural science turn, in large part, on whether we endorse a substance- or function-based picture of concept formation. Are basic concepts, like 'space', 'number', or 'cause', and more specific concepts, like 'matter', 'atom', 'irrational number', or 'point at infinity' formed on the basis of us 'copying' a substance? Or are they functions that mathematical and physical objects, in some sense, conform to? By Cassirer's lights, it is this fundamental disagreement about concept formation that gives rise to competing accounts of mathematical and scientific reality, truth, objectivity, and knowledge.

In a bit more detail, Cassirer traces the substance-based approach to concepts back to Aristotle and other 'empiricist' philosophers, like Berkeley and Mill.[4] Within this framework, Cassirer claims that concept formation is understood as a process of abstraction, according to which we form concepts by comparing several substances together, setting aside their differences, and abstracting out a property that they share in common (see SF Ch. 1). Insofar as the resulting concepts represent a generic characteristic or property of a substance, Cassirer dubs them 'substance-concepts' or 'thing-concepts' (*Dingebegriffe*). And he suggests that substance-concepts should be understood as 'abstract' universals, which neglect the specific difference between particulars and represent the generic features that particulars share in common (see SF 20–21/20).

By contrast, Cassirer aligns a function-based approach to concepts with the idea that concepts are 'functions'. Of course, one immediately wonders what Cassirer means by calling concepts 'functions'.

And it can be challenging to pin down exactly what notion of 'function' he has in mind, as the term takes on mathematical, logical, and organic connotations throughout *Substance and Function*. However, Cassirer's basic notion of a function involves a blend of Kantian and Russellian ideas.

In the Kantian vein, Cassirer is influenced by Kant's account of concepts as functions in the first *Critique*:

> concepts [rest] on functions. By a function . . ., I understand the unity of the action of ordering different representations under a common one. Concepts are therefore grounded on the spontaneity of thinking.
>
> (CPR A68/B93)

Kant, thus, characterizes a concept as a function insofar as it involves an activity by means of which we bring unity and order to representations. And in Kant's account, Cassirer sees an alternative to the more passive account of substance-concepts as 'copies' of what we are given.

Meanwhile, in the Russellian vein, in both "Kant and the Modern Mathematics" and the third volume of *The Philosophy of Symbolic Forms*, Cassirer explicitly aligns his approach to functions with Russell's analysis of propositional functions.[5] As Cassirer presents it, a propositional function comprises two components: a functional relation, $\emptyset$, and the variables, a, b, c, etc., that are ordered into a series through that relation (see PSFv3 363/358/301). Propositional functions thus have the form $\emptyset$(a, b, c . . .).

Combining these Kantian and Russellian ideas together, on Cassirer's view, a 'function-concept', or what he sometimes calls a 'relation-concept' (*Relationsbegriff*), involves a functional relation, which actively orders and unifies variables together into a series. For example, a mathematical concept, $\frac{n}{n+1}$, is a function-concept insofar as it involves a functional relation that actively orders numbers into a series, and which we can model as $\frac{n}{n+1}$ $(\frac{1}{2}, \frac{2}{3}, \frac{3}{4}, \frac{4}{5} \cdots)$ (PSFv3 364/359/312). Though a functional rendering of mathematical concepts is more familiar, on Cassirer's view, we can also think of empirical concepts along functional lines. For example, he claims that the empirical concept 'flying' involves a functional relation that actively orders things like birds, butterflies, and airplanes, into a series, and which we can model as flying (bird, butterfly, airplane . . .) (PSFv3 364/359/312).

According to Cassirer, unlike substance-concepts which are 'abstract' universals that we derive from particulars, function-concepts are 'concrete' universals by means of which we bring order and unity to particulars in the first place (SF 19–21/18–20). As he sometimes makes this point, a function-concept is a 'serial principle' that connects together particulars as a series:

> Here no insuperable gap can arise between the "universal" and "particular," for the universal itself has no other meaning and purpose than to present [*Darstellung*] and render possible the connection [*Zusammenhang*] and order of the particular itself. If we regard the particular as a *serial member* and the universal as a *serial principle*, it is at once clear that the two moments, without going over into each other and in any way being confused, still refer throughout in their capacity [*Leistung*] to each other.
>
> (SF 224/244, translation modified, emphasis in original)

Or, as he makes this point elsewhere, a function-concept is not 'blended with' particulars; it is a functional relation that provides a 'rule of progression' for particulars:

> That which binds the elements of the series $a, b, c, \ldots$ together is not itself a new element, that was factually blended with them, but it is the rule of progression, which remains the same, no matter in which member it is represented. The function $F(a, b)$, $F(b, c), \ldots$, which determines the sort of dependence between the successive members, is obviously not to be pointed out as itself a member of the series, which exists and develops according to it.
>
> (SF 17/17)

For example, as a function-concept, 'flying' is not a concept abstracted from birds, butterflies, and airplanes. It is, instead, a functional relation or rule, that 'renders possible the connection and order' of birds, butterflies, and planes as flying objects. And here we see the

Copernican thought that objects of cognition conform to functions of cognition at work.

It is in this dialectic between substance- and function-based theories of concepts that Cassirer situates his analysis of the facts of mathematics and natural science. And one of the main theses is that we can only offer an adequate account of these facts if we embrace a function-based, rather than a substance-based account of the concepts, cognition, and the objects of mathematical and scientific cognition.

Logical structuralism

With this substance-function framework in place, I shall now take up Cassirer's philosophy of mathematics. I begin with how he sets up this dialectic in the context of mathematics and then lay out the basic contours of logical structuralism, as the function-based view he seeks to vindicate.

According to Cassirer, the substance-based approach underwrites many different philosophical accounts of mathematics. Indeed, under this umbrella Cassirer includes ancient Greek views, which identify pure intuitive figures as the relevant substances; Berkeley's view, which identifies mental objects as the relevant substances; Mill's view, which identifies objects outside of us as the relevant substances; and Frege's and Russell's views, which identify logical classes as the relevant substances. The differences between these views notwithstanding, Cassirer categorizes them as substance-based views insofar as they treat mathematical concepts as ones we form by 'copying' substances through a process of abstraction.

Cassirer contrasts these substance-based accounts with a function-based account of mathematics, which I shall refer to as 'logical structuralism'. At the core of logical structuralism is the idea that mathematics is the science of ideal structures that are generated on the basis of functions or relations, which have a logical basis. Thus, on this view, rather than treating mathematical concepts as 'copies' of substances, mathematical concepts are understood in terms of functions or relations that generate mathematical structures.

In more detail, Cassirer presents logical structuralism as a position that has both a foundational and an ontological component. On the foundational point, logical structuralism is a logicist view, according to which logic is the foundation of mathematics. For Cassirer, the relevant logic is not first-order logic, but rather the 'logic of relations' of the sort Russell offers, which treats relations and functions as part of logic (see KMM 4/40, 7/43; SF 37/38; PSFv3 344–345/337/8; 293–294). And it is this logic of relations that Cassirer takes to be the foundation of the concepts, method, and objects of mathematics.

As a kind of logicism, logical structuralism thus differs from intuition-based theories of the foundations of mathematics, like Kant's. From the perspective of logical structuralism, Cassirer asserts,

> The intuition of pure time, upon which Kant based the con-cept of number, is indeed unnecessary. True, we think the members of the numerical series as an ordered sequence; but this sequence contains nothing of the concrete character of tem-poral succession. The "three" does not follow the "two" as light-ning the thunder . . . The lower number is "presupposed" by the higher number; but this . . . impl[ies] a pure relation of system-atic conceptual dependence.
>
> (SF 40/40)

As we see here, Cassirer thinks that although Kant is right to recog-nize that numerical series involve an order of progression, the logical structuralist faults Kant for thinking of this as an intuitive progression in time because it, in fact, is a logical progression. Thus, according to Cassirer, in order to account for the foundations of mathematics, the logical structuralist looks to logic rather than intuition.

However, Cassirer also distances logical structuralism from the logicism of Frege and Russell. As we saw above, Cassirer claims that Frege and Russell defend a substance-based version of logicism, which identifies logical classes as the 'substances' that we form mathematical concepts on the basis of.[6] By contrast, logical struc-turalism is a function-based view, according to which mathematical concepts are functions or relations, which generate mathematical structures in the first place.

This latter claim points toward the ontological aspect of logical structuralism. As Cassirer articulates it, logical structuralism is an account of mathematical objects, according to which they are not defined in terms of any intrinsic properties, but rather in terms of their place in ideal structures that are generated on the basis of functions or relations. Emphasizing the 'ideal' nature of mathematical objects on this view, he says,

> [in mathematics] a field of free and universal activity is disclosed, in which thought transcends [*hinauswächst*] all limits of the "given." The objects, which we consider and into whose objective nature we seek to penetrate, have only an ideal being; all the properties, which we predicate of them, flow exclusively from the law of their original construction.
>
> (SF 112/121)

And highlighting the 'relational' nature of mathematical objects on this view, he asserts, "The relational structure [*Relationsstruktur*] as such, not the absolute property of the elements constitutes the real object of mathematical investigation" (SF 93/98).

According to Cassirer, logical structuralism thus entails a revision of the traditional view of mathematics as the science of magnitudes, that is, of measurable substances or properties of substances, in favor of a view of mathematics as the science of relational structures:

> the task of mathematics does not consist in comparing, dividing or compounding given *magnitudes* [*Größen*], but rather in isolating the *generating relations* [*erzeugenden Relationen*] themselves, upon which all possible determination of magnitude rests, and determining their relationship to one another. The elements and all their derivatives appear as the result of certain original rules of connection [*Verknüpfung*], which are to be examined in their specific structure as well as in the character that results from the composition and interpenetration.
>
> (SF 95/101–102, translation modified,
> emphasis in original)

For the logical structuralist, then, the elements of mathematics are not to be understood in terms of the properties of given magnitudes, but rather in terms of their position within ideal structures that are generated on the basis of functions or relations. Thus, the 'task' of mathematics is not to be understood in terms of forming mathematical concepts that 'copy' given magnitudes, but rather in terms of forming mathematical concepts that represent the 'generating relations' or 'rules of connection' that generate these ideal structures.

To sum up, Cassirer conceives of logical structuralism as a theory of mathematics according to which the objects of mathematics are ideal structures that are generated on the basis of functions or relations, which have their source in logic. As we shall now see, he argues that in order to account for the 'new' facts of arithmetic and geometry, a substance-based view of mathematics is inadequate and that we should endorse logical structuralism instead.

Arithmetic

The 'new' facts that Cassirer orients his analysis of arithmetic around pertain to the admission of negative, imaginary, irrational, and transfinite numbers. However, on his view, in order to account for this fact, it is not enough to focus on these 'ideal' or 'extended' elements alone; one also has to account for their continuity with the natural numbers:

> For it is not enough that the new elements should prove equally justified with the old, in the sense that the two can enter into a combination [Verbindung] that is free from contradiction. . . . we show that the new elements not only are "adjoined" to the old ones as formations [Gebilde] of a different kind and origin but also are a systematically necessary unfolding of the old. And the proof of this interconnection can be undertaken only through the demonstration that there exists an originary logical kinship between the new and old elements. . . . It is from this point of view that the decisive achievement of the ideal elements must in the end be posited and from it that they must be understood and justified.
>
> (PSFv3 457–458/451–452/392, emphasis in
> original, translation modified)

Cassirer thus sets a further constraint on an adequate analysis of the 'new' facts of arithmetic: it must make sense of the continuity between the 'ideal' and 'extended' numbers and the natural numbers.

According to Cassirer, the inadequacy of a substance-based approach to the 'new' fact of modern arithmetic becomes readily apparent: there simply are no substances that our concepts of negative, imaginary, irrational, and transfinite numbers could serve as copies of. As he says about negative numbers, for example, "A negative substance, which would be at once being and not-being, would be a *contradictio in adjecto*" (SF 56/58). For this reason, Cassirer thinks that a function-based, rather than a substance-based approach is needed to account for these new mathematical concepts:

> The meaning of the generalized [i.e., ideal] concepts of number cannot be grasped as long as we try to indicate what they mean with regard to substances. . . . But the meaning at once becomes intelligible when we regard the concepts as expressing pure relations through which the connections [*Verhältnisse*] in a constructively produced series are governed.
>
> (SF 56/58)

In developing his function-based alternative, Cassirer appeals specifically to Dedekind and Dedekind's account of number along logical structuralist lines. By Cassirer's lights, Dedekind's logical structuralism makes sense not only of the irrational numbers, but also the continuity with the natural numbers. And Cassirer sees in Dedekind a model we should follow in accounting for the foundations and objects of arithmetic more generally.

In his analysis of Dedekind, Cassirer focuses primarily on the logical structuralism he sees at work in *What Are Numbers and What Are They For?* (*Was sind und was sollen die Zahlen?*) (1888). As Cassirer presents it, Dedekind pursues a logicist strategy that involves a "deduction of the concept of number, in its full import, from purely logical premises" (SF 35/36). In this deduction, Dedekind takes as his starting point the idea that,

> "what we do in counting a group or collection of things, we are led to consideration of the power of the mind to relate thing to thing, to let one thing correspond to another thing . . . a capacity in general without which thought is impossible. Upon this one, but absolutely inevitable foundation the whole science of number must be erected"
>
> (SF 36/36, quoting Dedekind)

The point of departure of Dedekind's deduction of the concept of number is thus the relational capacity of the mind that lies at the basis of thought. And while this might seem psychologistic, on Cassirer's reading, this relational capacity of the mind is not indexed to an individual psyche, but rather to the conditions of thought in general. So understood, what Dedekind's deduction of number really presupposes is the "general *logic of relations*" which represents the general conditions of relational thinking (SF 37/37).

Moreover, on Cassirer's analysis, Dedekind's position involves a structuralist understanding of numerical objects:

> The "things," which are spoken of in the . . . deduction, are not assumed as independent existences present anterior to any relation, but they gain their whole being, so far as it comes within the scope of the arithmetician, first in and with the relations, which are predicated of them. Such "things" are terms of relations, and as such can never be "given" in isolation but only in ideal community with each other.
>
> (SF 36/36)

Continuing in this vein, Cassirer says, on Dedekind's view, "the individual number represents nothing but . . . a particular place in a universal order structure [*universellen Ordnungsgefüge*]; and this structure determines the character and essential nature of the number" (PK 64–65/74). As we see here, on Cassirer's reading of Dedekind, numbers are not simply given as the substance-based view would have it, but rather exist as members of ideal structures that are generated on the basis of logical functions.

Cassirer goes on to examine how Dedekind offers a deduction of the concept of natural and irrational numbers within this framework of logical structuralism. With respect to natural numbers, Cassirer reconstructs Dedekind's deduction as follows:

> If, now, we consider a series which has a first member, and for which a certain law of progress has been established, of such a sort that to every member there belongs an immediate successor with which it is connected [*verknüpft*] by an unambiguous

transitive and asymmetrical relation, that remains throughout the whole series, then, in such a "progression," we have already grasped the real fundamental type with which arithmetic is concerned.

(SF 38/38, emphasis in original)

Cassirer takes Dedekind to logically deduce the natural number series by positing an initial member and then applying a transitive, asymmetrical relation to that member and its successor, the result being the successive series (1, 2, 3, 4, . . .). This series, Cassirer claims, is one that should be understood along structuralist lines as "a system of ideal objects whose whole content is exhausted in their mutual relations" and whose "'essence' . . . is completely expressed in their positions" (SF 39/40).

However, Cassirer points out, it is not just in the case of natural numbers that Dedekind uses this strategy: he extends it to his deduction of irrational numbers as well. As Cassirer articulates Dedekind's view, there is a "conceptual rule" on the basis of which the irrational numbers can be deduced: start with the rational numbers and then introduce a 'cut' a that divides the rational numbers into two classes, A and B (SF 61/63). Wherever this cut does not correspond to a rational number, there is an irrational number:

The "cuts" may be said to "be" numbers, since they form among themselves a strictly ordered manifold in which the relative position of the elements is determined according to a conceptual rule.

(SF 61/63)

As Cassirer sees it, Dedekind thus conceives of the irrational numbers as an ideal structure, a 'strictly ordered manifold', that is derived from a logical basis. And, on Cassirer's reader, what ultimately enabled Dedekind to establish the 'new' facts of irrational numbers was his logical structuralist recognition that our numerical concepts in general are not copies of substances we are given, but rather relations or functions on the basis of which ideal numerical structures are generated.

By Cassirer's lights, Dedekind's model of logical structuralism serves as an effective model for understanding the 'new' facts of arithmetic more generally:

> the key to a true understanding of the "ideal" formations [*Gebilde*] must be sought in the fact that the ideality by no means *begins* with them. . . . For there is no single truly mathematical concept that refers simply to pre-given and pre-encountered objects. . . . Here, it is always the positing of a general relation [*Relation*] that comes first, and it is only from its carrying out from all sides that the particular sphere of objects develops . . . Thus, essentially, the introduction of even-more-complex formations [*Gebilde*] only continues a process that was already begun and anticipated in the first "elements" of mathematics. . . . The same process is at work in both.
>
> (PSFv3 460/454–455/395, emphasis in original)

According to Cassirer, in order to make sense of the concepts of 'ideal' numbers more generally, we must recognize, as Dedekind did, that these concepts, like all numerical concepts, serve as the logical basis on which numerical series are generated. To be sure, Cassirer acknowledges that there are some differences between the elementary and 'extended' numerical series. Whereas in the case of the natural numbers, the logical function is applied to a simple element and only includes relatively simple elements, in the case of the ideal numbers, the logical function is applied to the series of rational numbers, the result being a more 'complex' and 'inclusive' numerical series (SF 61/63). The ideal numbers will thus involve more complex, inclusive ideal structures than the natural numbers do. However, Cassirer regards this as a difference in degree, not kind:

> ordinal numbers . . . signify nothing but positions in an ordered series. . . . Once this fundamental character of number is recognized . . . there are no further difficulties about the exten-sion of the domain of numbers. . . . The task . . . is simply that of advancing from a system of relatively simple relationships to more complex systems of relations.
>
> (PK 68/78)

For Cassirer, then, the logical structuralism that underwrites Dedekind's deduction of the natural and irrational numbers provides a unified picture for understanding the nature of numerical concepts, the foundations of arithmetic, and numbers more generally.

Extending Dedekind's model to the other 'new' facts, Cassirer appeals to logical structuralism in order to make sense of the concepts of negative, imaginary, and transfinite numbers. In the case of negative numbers, for example, Cassirer argues that their deduction depends on applying an "inverse relation" to the series of positive numbers (see SF 56/58). Likewise, with imaginary numbers, Cassirer claims that Gauss deduces them by applying the successor relation and the inverse of the successor relation to "a *series* of series," i.e., a series of series of natural numbers, and thereby introducing the "new unity (+i, -i)" (SF 56/58). Meanwhile, with respect to transfinite numbers, Cassirer asserts that Cantor derives this series by logically positing ω as the limit of the natural number series and then generating a new series by applying the "second principle of generation" to ω (see SF 65–66/67–69). In each instance, Cassirer maintains that we have a numerical series that should be understood as a complex ideal structure that is generated on the basis of a numerical concept, qua a logical function that generates a numerical series. What is more, in keeping with the Dedekind model, Cassirer claims that this logical derivation of the negative, imaginary, and transfinite numbers should be understood as the 'continuous unfolding' of the logical process at work in the natural numbers: "the new structures [*Gebilde*] are not added to the number system from without but grow out of the continuous unfolding of the fundamental logical function that was effective in the first beginnings of the system" (SF 67/69, translation modified).

On Cassirer's view, then, reflection on the 'new' facts of modern arithmetic, as well as considerations about their continuity with the 'old' facts, give us reason to endorse a logical structuralist view of arithmetic. Whether we consider our concepts of elementary or 'ideal' numbers, instead of regarding them as 'copies' of substances, we should understand them as function-concepts, which have their source in logic, on the basis of which mathematical structures are generated.

Geometry

Turning now to Cassirer's analysis of geometry, he takes as his starting point the 'new' facts pertaining to developments in the 19th century, including non-Euclidean geometry, projective geometry, and the application of group theory to geometry in Klein's Erlangen Program. As was the case with arithmetic, on Cassirer's view, an adequate analysis of these facts not only accounts for the conditions that make the new versions of geometry possible, but also explains the continuity between the new and old geometrical systems.

Cassirer, once again, proceeds by arguing that a substance-based view cannot properly account for these modern advances in geometry given the role that 'ideal' elements play. On a traditional substance-based view, we form geometrical concepts by 'copying' the figures in space we are given through intuition. However, as Cassirer points out, concepts of non-Euclidean and projective geometry, like a point or line at infinity and imaginary points, transcend "the limits of possible sensuous presentation [*sinnlichen Darstellbarkeit*]" (SF 78/82, translation modified). If we are to make sense of these new elements, Cassirer claims we must look to a function-, rather than a substance-based view.

In defending his function-based alternative, Cassirer extends the model of logical structuralism operative in Dedekind's arithmetic to geometry. To this end, he claims that the foundation of geometry is not intuition, but rather functions or relations that have a basis in logic, and that geometrical objects are not given in intuition, but are ideal structures that are generated on the basis of those functions or relations.

In order to make his case for logical structuralism, Cassirer connects together two developments in geometry in the 19th century: the invariant-based approach of Poncelet and Klein and the axiomatic approach of Hilbert.[7] In both of these developments, Cassirer sees a trend toward logical structuralism similar to what we find in Dedekind's arithmetic. And he thus takes these developments to give us reason to endorse the logical structuralist view not only of the 'new' facts of geometry, but of geometry as a whole.

Beginning with the invariant-based approaches, Poncelet's method for projective geometry is a method for investigating the geometrical properties of figures that remain invariant across transformations. By Cassirer's lights, this method calls for a shift away from intuition toward a 'rational deduction':

> The value of the new method [of Poncelet] is found in that through it geometrical deduction can proceed entirely unhindered, in that, without being narrowed to the limits of possible sensuous presentation, it especially takes account of imaginary and infinitely distant elements that possess no individual geometrical "existence," and thus first attains completeness of rational deduction.
>
> (SF 78/82, translation modified)

Indeed, Cassirer treats this 'rational deduction' as a geometrical counterpart of Dedekind's rational deduction of number:

> As in the case of number we start from an original unit [ursprüngliche Einheitssetzung] from which, by a certain generating relation, the totality of the members is evolved in fixed order, so here we first postulate a plurality of points and a certain relation of position between them, and in this beginning a principle is discovered from the various applications of which issue the totality of possible spatial constructions. In this connection, projective geometry has with justice been said to be the universal "a priori" science of space, which is to be placed beside arithmetic in deductive rigor and purity.
>
> (SF 88/93)

On Cassirer's view, then, the projective method is, in essence, like Dedekind's method in arithmetic: it proceeds 'deductively' by applying a 'generating relation' to an 'original unit'.

More specifically, Cassirer asserts that in Poncelet's projective geometry, "the particular form we are studying [is] ... merely ... a starting point, from which to deduce by a certain rule of variation a whole system of possible forms [Gestaltungen]" (SF 80/84, emphasis in original). Cassirer specifies this 'rule of variation' as a rule for "continuous transformation" (SF 80/85). And he claims that, for Poncelet, the 'true geometrical object' is the 'fundamental relations' that remain 'invariant' across continuous transformations:

> The fundamental relations, which characterize this system, and which must be equally satisfied in each particular form, constitute in their totality the true geometrical object [Objekt]. What the geometrician considers is not so much the properties of a given figure [Figur] as the net-work [Netz] of *correlations* [Korrelationen] in which it stands with other allied structures [Bildungen]. We say that a definite spatial form [Gestaltung] is correlative to another when it is deducible from the latter by a continuous transformation of one or more of its elements of position: yet in which the assumption holds that certain fundamental spatial relations, which are to be regarded as the general conditions of the system, remain unchanged. The force and conclusiveness of geometrical *proof* always rests then in the *invariants* of the system, not in what is peculiar to the individual members as such.
>
> (SF 80/84–85, emphasis in original)[8]

Thus, for Poncelet, the object of geometry is not a figure that is given to us in intuition; it is the invariant relations or structures that remain constant across transformation, which are determined through rules of transformation.

On Cassirer's reading, in Poncelet's invariant-based approach we find a geometrical analogue of Dedekind's logical structuralism: the geometrical method is a deductive, rather than intuitive method; geometrical concepts are function-concepts, i.e., rules for transformation, through which geometrical structures are generated; and geometrical objects are structures that are generated on the basis of rules of continuous transformations. What is more, like Dedekind, Cassirer argues that it is because Poncelet endorses this framework that he can offer an account of the continuity between elementary and ideal elements of geometry. As Cassirer glosses Poncelet's position, "real and imaginary elements are essentially similar; for the latter also are the expression of perfectly valid and true geometrical relations" (SF 83/88). Moreover, Cassirer maintains that insofar as

Poncelet regards real and ideal elements as "expressions [*Ausprägungen*] of *one and the same concept*," he treats them as logically and systematically connected in the way that natural and 'ideal' numbers are (SF 82/87, emphasis in original). Indeed, Cassirer suggests that, for Poncelet, it is precisely on account of their systematic connection to the real elements that the ideal elements can shed light on the real elements:

> the imaginary intermediate members always serve to make possible insight into the *connection* [*Verknüpfung*] of real geometrical forms [*Gestaltungen*], which without this mediation would stand opposed as heterogeneous and unrelated.
>
> (SF 83/88, emphasis in original)

From Cassirer's perspective, one of the merits of Poncelet's invariant-based approach is that it paves the way for an understanding of the continuity between real and ideal elements of geometry.

Turning to Klein, Cassirer regards Klein's Erlangen Program as a generalization of Poncelet's invariant-based approach to projective geometry to all geometries, Euclidean, affine, projective, and the like. For Klein, every geometry can be understood in terms of the following principle, "'*Given a manifold, and in it a transformation group; one should investigate the figures belonging to the manifold with a view to finding the properties that are not changed by the transformation of the group*'" (PK 31/36, quoting Klein). That is to say, every geometry investigates the properties of geometrical figures that remain invariant across transformation, the difference between them being a matter of which group of transformations one selects. For example, the transformation rules in Euclidean geometry involve metric requirements, which are dropped in affine and projective geometry. As Cassirer summarizes Klein's view, "Every geometry in its general concept and aim is a *theory of invariants* [*Invariententheorie*] with respect to a certain group, and the special nature of each depends on the choice of the group" (PK 31/35, emphasis in original).

According to Cassirer, Klein is thus not only able to offer a unified account of geometry as a 'theory of invariants', but also a precise 'hierarchy' of the different geometries (PK 32/37). For example, in this hierarchy, projective geometry is 'above' Euclidean geometry, in the sense that, "the group of projective transformations contains as a part of itself the 'principal' group upon which Euclidean geometry rests" (PK 32–33/37).

As Cassirer sees it, Klein thus broadens Poncelet's insight that projective geometry is the study of the relations that remain invariant across transformations to all geometry. And given Cassirer's understanding of this insight in logical structuralist terms, he takes Klein's Erlangen Program to be further evidence in favor of logical structuralism in geometry. Indeed, on Cassirer's reading, what the Erlangen Program reveals is that logical structuralism provides a unified account of the foundations, objects, and concepts of geometry as a whole.

However, according to Cassirer, the trend toward logical structuralism does not just underwrite Poncelet and Klein's invariant-based approach, it also animates the axiomatic approach endorsed by Hilbert. In *Foundations of Geometry* (1899), Hilbert argues that geometry has its basis not in what can be given in intuition, but rather on the basis of geometrical axioms. As Cassirer presents Hilbert's view,

> In contrast to the Euclidean definitions, which take the concepts of the point or the straight line as immediate data of intuition . . ., the nature of the original geometrical objects is here exclusively defined by the conditions to which they are subordinated. The beginning consists of a certain group of axioms, which we assume . . . From these rules of connection [*Zusammenhörigkeit*] . . . follow all the properties of the elements. The point and straight line signify nothing but formations [*Gebilde*] which stand in certain relations with others of their kind, as these relations are defined by a certain group of axioms. . . . In this sense, Hilbert's geometry has been correctly called a pure theory of relations [*Beziehungslehre*].
> (SF 93–94/99–100, translation modified)

On Hilbert's view, it is axioms, rather than intuition, which serves as the basis of elementary objects, like points and lines, and elementary relations between those objects. Insofar as he conceives of these axioms in logical terms, he thus embraces a logicist account of the foundations of Euclidean geometry. And insofar as he conceives of geometrical objects, like points and lines, as ones that are defined through these relations, he is committed to a structuralist view of them. Thus, instead of treating Euclidean geometry as the science of the space we are given in intuition, Cassirer takes Hilbert to embrace the logical structuralist view, according to which it is the science of structures that are generated on the basis of axioms, which have their seat in logic rather than intuition.

In his discussion of the 'new' facts of modern geometry, Cassirer thus appeals to Poncelet's and Klein's invariant-based approach and Hilbert's axiomatic approach as evidence in favor of logical structuralism. For what these approaches offer is a unified account of geometry, across its 'new' and 'old' instantiations as a science of geometrical structures, which are generated on the basis of functions or rules that have their source in logic, not in intuition. This account not only makes sense of the formation of new geometrical concepts, which represent 'ideal' elements, but also the continuity between the 'ideal' and elementary aspects of geometry. Thus, in these developments in geometry, Cassirer thus sees a geometrical manifestation of the kind of logical structuralism that Dedekind defends in geometry. Like arithmetic, geometry has its foundation in a logic of functions or relations; like numbers, geometrical objects are to be understood as structures that are generated on the basis of those functions or relations; and like numerical concepts, geometrical concepts are to be understood not as copies of a substance given in intuition, but in terms of these functions or relations. Thus, as was the case in arithmetic, Cassirer concludes that a careful examination of trends in geometry itself gives us reason to reject a structure-based view in favor of logical structuralism, as a function-based view that does better justice not only to the 'new' facts of geometry, but to the foundations and objects of geometry more generally.

Intuition and mathematics

As I have presented it so far, Cassirer's defense of logical structuralism appears to leave little room for intuition to play any significant role in mathematics. And he, indeed, argues that the logicist account of the foundations of mathematics is confirmed by the developments in modern mathematics itself: "modern mathematics . . . [has] turned mathematics more and more into a 'hypothetical-deductive system,' whose truth-value is grounded purely in its inner logical coherence and consistency, not in any material, intuitive statements" (PSFv3 427/418–419/363–364). Yet, as Cassirer acknowledges, it is precisely this logicist trajectory that was called into question by the so-called 'foundational crisis' in mathematics in the early 20th century (see PSFv3 Pt. 3, Ch. 4.2).[9] This crisis was precipitated by considerations about the paradoxes facing set theory, including those of Frege and Dedekind, as mathematicians and philosophers began to question whether a valid mathematical object could be determined solely on the basis of logical criteria or whether extra-logical criteria were needed. Two alternatives to logicism thus emerged. Intuitionists, like Henri Poincaré, L.E.J. Brouwer, and Hermann Weyl, argued that the foundations of mathematics should be traced back to the constructive activity of an idealized, though finite mathematician. Meanwhile, formalists, like Hilbert, claimed that mathematics is grounded in the formal game we play with mathematical symbols.

What is of particular interest to me in Cassirer's analysis of the foundational crisis is his appraisal of intuitionism. Though one might have expected him to reject intuitionism in favor of logicism, this is not exactly what he does. To be sure, he objects to the intuitionist claim that intuition is the foundation of mathematics, and he is also highly critical of the particularly psychologistic version of intuitionism defended by Brouwer. Nevertheless, Cassirer asserts intuitionism gets something right that the logicism of Frege and Russell does not:

> Those trying to establish a purely logical foundation of number . . . lost sight almost entirely of the theoretical problem. Number was raised so high above the world of intuition that it was hard to understand how . . . it could be related to the empirical world. . . . It was "intuitionism" that first succeeded in constructing a bridge.
>
> (PK 79/91, translation modified)[10]

Here, Cassirer objects that Frege and Russell lose sight of the importance of the application of number to the empirical world through natural science and he praises intuitionism for having given a positive account of it. These laudatory comments, in turn, point toward the fact that although Cassirer endorses a logicist account of the foundations of mathematics, he nevertheless accords intuition some role in relation to mathematics. To clarify Cassirer's account of the relationship between intuition and mathematics, I shall consider three ways he thinks intuition can make a positive contribution.

In the first place, he claims that intuition is often the occasion for our mathematical procedures. The intuition of concrete spatial figures, such as, five lines in a row or a triangle on a chalkboard, can serve as the starting point of arithmetical or geometrical investigations. However, Cassirer is clear that this does not mean that these intuitive figures are, therefore, the foundation of mathematical concepts. Rather, he argues we must distinguish between "the psychological *beginning*" and "logical *ground*" of the mathematical procedure:

> It is true that, in the psychological sense, we can only present the meaning of a certain relation to ourselves in connection with some *given terms*, that serve as its "foundations" [*Fundamente*]. But these terms, which we owe to sensuous intuition, have no absolute, but rather a changeable existence. We take them only as hypothetical starting-points; but we look for all closer *determination* from their successive insertion into various relational complexes. It is by this intellectual process that the provisional content first becomes a fixed logical object. The law of connection, therefore, signifies the real "first by nature", while the elements in their apparent absoluteness signify only a "first for us."
>
> (SF 94/100, emphasis in original)

By putting 'foundations' in quotes here, Cassirer means to indicate that although intuitive elements can serve as the psychological starting point of mathematics, the true foundation of mathematics is logical in nature.

Second, Cassirer maintains that we rely on intuition in order to represent mathematical concepts to ourselves. Making this point in "Kant and the Modern Mathematics," he claims, intuition is "not the *origin* [*Ursprung*] of logical and mathematical principles, instead it . . . only brings them to concrete *presentation* [*Darstellung*]"

(KMM 67/33). Or in The Philosophy of Symbolic Forms, he says, "Mathematics no longer appealed to intuition as a positive means of proof and grounding; rather, it used intuition only to give a concrete *representation* [*Repräsentation*] to the general relational interconnections that it constructed in pure thinking" (PSFv3 427/419/364, emphasis in original). In this vein, Cassirer takes his cue from Leibniz. On Cassirer's interpretation, even though Leibniz thinks that, "[n]o mathematical content arises as such from sensibility," but rather has its foundation in "certain relations that prevail among pure ideas," Leibniz nevertheless recognizes that in order for us to "master and retain" these concepts we must "mak[e] them tangible in images, in symbols" (PSFv3 424/414–415/360–361, PSFv3 424/414/360). Indeed, Cassirer suggests that on Leibniz's view, "Our finite understanding is and will remain an understanding in need of images" (PSFv3 424/415/361). Cassirer endorses this Leibnizian idea, arguing that even though intuition is not the foundation of mathematics, we need to represent mathematical concepts to ourselves through intuitive symbols in order to make them comprehensible to ourselves.

Third, as I alluded to above, Cassirer argues that intuition has a role to play with regard to the empirical application of mathematics in natural science. Returning to his criticism of Frege and Russell along these lines, Cassirer says,

> Those trying to establish a purely logical foundation of number, constructing it from purely "logical constants," had indeed attained considerable formal rigor, but in so doing had lost sight almost entirely of the cognitive-theoretical [*erkenntnistheoretische*] problem. Number was raised so high above the world of intuition that it was hard to understand how, without prejudice to its "purity," it could be related to the empirical world; indeed, even become a specific principle of cognition therein.
>
> (PK 79/91, translation modified)

Putting this point in Kantian terms Cassirer claims that Frege and Russell treat mathematics as related solely to 'general' or 'formal' logic, and thereby neglect the role mathematics plays in 'transcendental' logic, that is, the logic that concerns our cognition of objects (KMM 77–78/44–45, see CPR A55/B79–A57/B82). However, on

Cassirer's view, appreciating the role mathematics plays in our cognition of objects in natural science is essential to understanding the nature of mathematics more generally.

For Cassirer, then, a proper philosophy of mathematics has two tasks. First, there is the "purely logical" task of clarifying the relations and functions that serve as the ground of the mathematical method and mathematical objects (PK 79/91). And it is this task that Cassirer seeks to fulfill with his defense of logical structuralism. Second, there is a task "where logicism leaves off," viz., a critical task that involves examining the applicability of mathematics and the contribution it makes to the scientific cognition of nature (KMM 77/44). This latter task, Cassirer claims, is one that requires that we broaden our perspective beyond the realm of pure mathematics and look to the role that mathematics plays in the "logic of the exact science of nature" (PK 80/92). And, on Cassirer's view, unless we consider how mathematics contributes to the "whole *objectivization process of cognition*" in natural science, our understanding of the nature of mathematics remains incomplete (PSFv3 448/442/384, emphasis in original).

Cassirer buttresses his insistence on the importance of taking questions of application into account by pointing toward a development in modern accounts of mathematical axioms. According to Cassirer, whereas in antiquity, mathematical axioms were conceived along Platonic lines, "enthroned in solitary grandeur above the world of the senses," in modern mathematics, mathematical axioms have come to be understood in terms of their 'readiness' for empirical application:

> However little [mathematical thinking] is derived from the sense world it still displays a constant readiness for sense material, and here the modern concept of axioms differs characteristically from the ancient. . . . They are . . . proposals [*Ansätze*] for thought that make it ready for application—schemata, which thought projects and which must be so broadly and inclusively construed as to be open to every concrete application that one wishes to make of them in cognition. . . . According to modern views a system of axioms is . . . to be thought of . . . as a "logical blank form of the possible sciences".
>
> (PK 45–46/52, translation modified)

Elaborating on this idea in his discussion of geometry in *Substance and Function*, Cassirer claims that insofar as geometrical systems are a "whole of relations [*Ganzes von Beziehungen*] . . . they make no final determination of the character of the individual members, which enter into these relations" and that "a field of *application* for the abstract propositions is indicated" (SF 110/118, translation modified). On Cassirer's view, insofar as mathematical systems are relational systems that make no decisions about the nature of their relata, they remain open to applications in which the character of those relata are determined through the empirical research of natural science.

According to Cassirer, appreciating the significance of the question of applicability for understanding the nature of mathematics more generally points toward the third role that intuition plays in relation to mathematics. From Cassirer's perspective, although intuition does not have a role to play in the foundation of mathematics, it does in the selection of which mathematical theories to use in empirical application. On this issue, Cassirer appeals to Poincaré's idea that intuition, as part of experience, can serve as a 'principle of selection' in the application of geometry, "we see in experience a *principle of selection* [*Auswahlprinzip*] for the application of geometry, even though experience does not thereby become its genuine *justificatory ground* [*Rechtsgrund*], its 'quid juris'" (PK 45/51, emphasis in original, translation modified). Sanctioning this sort of view in *Substance and Function*, here in his own voice, he says, "The role, which we can still ascribe to *experience*, does not lie in *founding* [*Begründung*] the particular [mathematical] systems, but in the *selection* that we have to make among them" (SF 106/114, emphasis in original). Insofar as intuition is part of experience, on Cassirer's view, it does not found mathematical systems, but can nevertheless play a role in guiding us toward selecting which mathematical systems to use for empirical application in natural science.

Ultimately, in response to the foundational crisis in mathematics, although Cassirer remains firm in the conviction that mathematical relations and the structural objects they generate have a logical ground, he acknowledges various roles for intuition to play in relation to mathematics: it can serve as the psychological beginning of mathematical cognition, it can help us represent mathematical concepts to ourselves, and it can contribute to the empirical application of mathematics in scientific cognition. And the last contribution is one that Cassirer underlines, in particular, as one we need to acknowledge in order to do justice to mathematics as a whole.

Summary

My goal in this chapter was to begin exploring the Neo-Kantian philosophy of mathematics and natural science that Cassirer defends and his effort to execute the transcendental method with respect to the revolutionary 'new' facts that emerged in these fields in the 19th and 20th centuries. After discussing the substance-function framework that Cassirer develops to this end, I turned to his philosophy of mathematics and the case he makes for why the facts of modern arithmetic and geometry demand a function-based, rather than a substance-based approach. I laid out Cassirer's preferred function-based approach, viz., 'logical structuralism', according to which mathematics has its basis in functions of relations that belong to logic and mathematical objects are ideal structures generated on the basis of those functions or relations. I then turned to his argument that the developments in modern arithmetic and geometry give us reason to endorse logical structuralism. To this end, I considered how Cassirer draws on Dedekind's model of logical structuralism in order to account for the method, concepts, and objects of arithmetic. I next discussed his argument that the invariant-based approach of Poncelet and Klein and the axiomatic approach of Hilbert should be understood as a geometrical analogue of Dedekind's logical structuralism in arithmetic.

Though Cassirer thus endorses a logicist account of the objects and foundations of mathematics, I also highlighted the ways in which he takes his view to differ from the logicism of Frege and Russell. Although Cassirer defends a logicist account of the foundation of mathematics, he criticizes Frege and Russell for neglecting the role that intuition plays in mathematical thinking and for failing to acknowledge that in order to do justice to mathematics, we must attend to the question of its applicability to the empirical world in natural science. And I concluded with a discussion of the positive roles that Cassirer attributes to intuition not just in the empirical application of mathematics, but also in occasioning mathematical thinking and enabling us to represent mathematical concepts to ourselves.

Notes

1 In his philosophy of natural science, although Cassirer discusses historical developments in biology in PK, Pt. 2 and addresses it in passing in Ch. 10 of EM, his most sustained efforts are directed towards the 'mathematical' sciences of physics and chemistry. By 'natural science' in what follows, I thus mean 'mathematical' natural science.

2 Translations of KMM ("Kant und die moderne Mathematik" ["Kant and Modern Mathematics"]) are my own.

3 In his philosophy of mathematics, Cassirer uses 'arithmetic' to refer to the mathematical study of numbers and basic operations with numbers, like addition, subtraction, etc. He also addresses the topic of number in his philosophy of culture, as a category that is operative in other cultural regions, like myth and religion (see Chapters Five and Six).

4 Cassirer devotes Chapter One of Substance and Function to a critical discussion of substance-based theory of concepts and concept formation in general. After discussing the origins of this view in Aristotle and its development in Berkeley and Mill, Cassirer raises the general objection that the substance-based theorist offers a circular account of concept formation through 'abstraction' (see, e.g., SF Ch. 1, §§2–3). He also levels specific objections against particular substance-based views: see, e.g., his discussion of Frege's objections to Mill's view of number (SF 28–30/28–30), his presentation of Descartes's criticism of the synthetic view of geometry (SF 70–73/73–75), and his own criticism of the logicist view of Frege and Russell for neglecting the applicability of number (KMM, PK 79/91).

5 Cassirer directs us toward Russell's analysis of functions along propositional lines in "Sur la relation des mathématique à la logistique" (see KMM 43 n12/7 n3) and Principles of Mathematics (see PSFv3 346–347/339–340/301).

6 I also discuss his criticism of Frege and Russell's logicism for neglecting questions of the empirical applicability of mathematics at the end of this chapter.

7 See Ihmig (1997), (1999) for a discussion of the originality of Cassirer's insight into the connection between Klein's Erlangen Program and Hilbert's axiomatic approach.

8 See Chapter Four for a discussion of the role 'invariant relations' play in Cassirer's philosophy of natural science.

9 Cassirer offers a longer historical introduction to the foundational crisis that traces it back to Leibniz and Kant in PSFv3 Pt 3, Ch. 4, §1.

10 Cassirer also praises intuitionism for insisting on the 'primacy of relation' over and against the substance-based logicism of Frege and Russell: "It is to the merit of 'intuitionism' over and against all such [logicist] theories that it has restored the *primacy of the relation* and has brought about its basic recognition" (PSFv3 440/433/376, emph. in original).

Further reading

Biagioli, Francesca (2016). *Space, Number, and Geometry from Helmholtz to Cassirer.* Archimedes 46 Series. Springer. A survey of the developments in Neo-Kantianism, from Helmholtz through Cassirer, pertaining to arithmetic and geometry.

——(2020). "Ernst Cassirer's transcendental account of mathematical reasoning." *Studies in History and Philosophy of Science* 79: 30–40. A discussion of how Cassirer attempts to reconcile Kantian commitments with modern mathematics in his account of mathematical reasoning.

Heis, Jeremy (2010). ""Critical Philosophy Begins at the Very Point Where Logicist Leaves Off": Cassirer's Response to Frege and Russell." *Perspectives on Science* 18.4: 383–408. A discussion of the features of Frege and Russell's logicism that Cassirer is both sympathetic to and critical of.

——(2011). "Ernst Cassirer's Neo-Kantian Philosophy of Geometry." *British Journal for the History of Philosophy* 19.4: 759–794. A discussion of Cassirer's philosophy of geometry, which emphasizes its Dedekind-inspired logicist and structuralist themes and the issue of applicability.

——(2014). "Ernst Cassirer's *Substanzbegriff und Funktionsbegriff.*" HOPOS 4: 241–270. A discussion of Cassirer's overarching motivations in *Substance and Function* regarding the topic of concepts and concept formation and how this bears on his philosophy of mathematics.

Ihmig, Karol-Nobert (1997). *Cassirers Invariantentheorie der Erfahrung und seine Rezeption des 'Erlangers Programms'.* Hamburg: Meiner. A full-length study of the relevance of Klein's Erlangen Program for Cassirer's philosophy of mathematics and natural science.

——(1999). "Ernst Cassirer and the Structural Conception of Objects in Modern Science: The Importance of the 'Erlanger Programm'." *Science in Context* 12(4): 513–529. A briefer discussion of the same topic.

Mormann, Thomas (2008). "Idealization in Cassirer's Philosophy of Mathematics." *Philosophia Mathematica* 16: 151–181. A discussion of the role that idealization plays in Cassirer's account of mathematics and natural science.

Reck, Erich (2020). "Cassirer's Reception of Dedekind and the Structuralist Transformation of Mathematics." In *The Pre-History of Mathematical Structuralism*, edited by Erich Reck and Georg Schiemer. Oxford: Oxford University Press. A discussion of Dedekind's influence on Cassirer.

Reck, Erich and Keller, Pierre (forthcoming). "From Dedekind to Cassirer: Logicism and the Kantian Heritage." In *Kant's Philosophy of Mathematics, Vol. II: Reception and Development After Kant*, edited by Ofra Rechter and Carl Posy. Cambridge: Cambridge University Press. A discussion of Dedekind's influence on Cassirer.

Richardson, Alan (2016). "On Making Philosophy Functional: Ernst Cassirer's *Substanzbegriff und Funktionsbegriff*." In *Ten Neglected Classics of Philosophy*, edited by Eric Schliesser, 177–194. Oxford: Oxford University Press. A brief overview of the main points of *Substance and Function* and its continued value for understanding the origins of analytic philosophy.

Schiemer, Georg (2018). "Cassirer and the Structural Turn in Modern Geometry." *Journal of the History of Analytic Philosophy*, Special Issue: Method, Science, and Mathematics: Neo-Kantianism and Early Analytic Philosophy, edited by Lydia Patton and Scott Edgar, 6(3): 183–212. A discussion of structuralist themes in Cassirer's account of geometry in *Substance and Function*.

Yap, Audrey (2017). "Dedekind and Cassirer on Mathematical Concept Formation." *Philosophia Mathematica* 25(3): 369–389. An anti-psychologistic reading of Dedekind's arithmetic in light of Cassirer's interpretation of it.

Four
Philosophy of natural science

Introduction

During Cassirer's lifetime, the landscape of (mathematical) natural science shifted in revolutionary ways.[1] Einstein's theories of special and general relativity appeared in 1905 and 1915, respectively; Niels Bohr introduced his model of the atom in 1913; and much of the groundbreaking work in quantum mechanics by Bohr, Louis de Broglie, Erwin Schrödinger, and Werner Heisenberg was published in the 1920s. Cassirer thus witnessed the 'fact' of modern physics and chemistry change in unprecedented ways and his philosophy of natural science represents his attempt to keep pace with these remarkable developments.[2] As a result, his writings on natural science have a particularly dynamic character. In *Substance and Function* (1910) he takes classical mechanics and chemistry as his starting point. He then shifts his focus to relativity in *Einstein's Theory of Relativity* (1921), becoming one of the first philosophers to do so. In the third volume of *The Philosophy of Symbolic Forms* (1929), he addresses both relativity and quantum mechanics. And he offers a book-length study of quantum mechanics in *Determinism and Indeterminism in Modern Physics* (1936). However, in spite of the ever-increasing scope of his philosophy of natural science, Cassirer remains committed throughout to defending a Neo-Kantian approach, grounded in the transcendental method, critical idealism, and a critical theory of cognition.

My aim in this chapter is to examine how Cassirer uses this Neo-Kantian framework to account for the possibility of the fact of modern natural science, especially in light of its radical changes in the 20th century. I begin in the next section by returning to Cassirer's substance-function dialectic and his general argument for why a function-based approach to natural science is to be preferred over a substance-based approach. As we shall see, central to Cassirer's argument is the claim that even the most basic practice of natural science, viz., measurement, depends on function-concepts. I then turn to the details of Cassirer's own version of the function-based view, starting with an account of how he deploys critical idealism and the critical theory of cognition in his account of natural science. Next I take up Cassirer's methodological remarks about natural science. In addition to analyzing the method of natural science itself, Cassirer offers an account of the method transcendental philosophers should use in relation to natural science. He, indeed, treats the natural scientist and transcendental philosopher as engaged in a shared project, viz., searching for 'invariant relations' that govern physical phenomena and scientific experience—the former pursuing invariant relations of the physical type and the latter relations of the *a priori* type. Lastly I address his analysis of the progress of natural science, concentrating specifically on his argument for the continuity of this progress even across radical paradigm shifts.

Substance and function, redux

In order to enter into Cassirer's philosophy of natural science, it will be helpful to begin, as he does, by situating it within his substance-function distinction. Recall that, on Cassirer's view, a substance-based view is one that adheres to the two basic presuppositions: a substance-based ontology and the copy theory of cognition. Meanwhile a function-based view is one that is grounded in a Kantian commitment to critical idealism, according to which the objects of cognition conform to the functions of cognition, and the critical theory of cognition, according to which the aim of epistemology is an analysis of these functions. As was the case with his account of mathematics, Cassirer argues that we can do full justice to the facts of natural science only if we adhere to a function-based, rather than a substance-based approach.

However, Cassirer acknowledges that the prospects for defending a function-based approach to natural science might seem dimmer than defending a function-based approach to mathematics. Given the more ideal orientation of mathematics, Cassirer thinks it is easier to motivate the thought that mathematical objects conform to mathematical functions. By contrast, when we think of natural science, it is tempting to think that the substance-based view gets it right: the goal is to develop scientific concepts and theories that provide faithful 'descriptions' that 'correspond' to the mind-independent world around us (see SF 112–113/121–122, 213/233; PSFv3 476–477/468–469/406–407). The thought that our physical world and the objects in it conform to functions seems to mischaracterize the nature of our scientific endeavors.

In spite of this *prima facie* concern, Cassirer argues that if we look at the actual practice of physics and chemistry, then it will emerge that the function-based view is to be preferred. In order to make his case, Cassirer begins by looking at what is arguably the most fundamental practice of natural science, viz., measurement (see SF Chap. 4, ETR Chap. 1, and DI Chap. 3). By Cassirer's lights, regardless of whether one endorses a substance- or function-based view, all parties should agree that the mathematical determination of objects through measurement is the bedrock of *mathematical* natural science. Making this point about physics, he says, "All physics considers phenomena under the standpoint and presupposition of their measurability. It seeks to resolve the structure [*Gefüge*] of being and process [*Geschehens*] ultimately into a pure structure [*Gefüge*] or order of numbers" (ETR 448/114).[3] Hence his endorsement of Max Planck's claim that in order for something to count as an "object for physics," it must be "something measurable" (ETR 357/8). And Cassirer takes the idea that an 'object for' natural science is 'something measurable' to be a basic commitment for mathematical natural science as such.

However, according to Cassirer, if we examine the practice of measurement, we shall find that the substance-based view is inadequate. To this end, he makes the case that the facts of measurement are not simple givens that we attempt to 'describe'; the facts of measurement are generated on the basis of functions.

In developing this argument, Cassirer claims that measurement is a theory-laden process:

> it is erroneous to regard measurement itself as a purely empirical procedure, which could be carried out by mere perception

and its means . . . for the numbered and measured phenomenon is not a self-evident, immediately certain and given starting-point, but the result of certain conceptual operations, which must be traced in detail.

(SF 141/153)

Continuing in this vein, he says, "the simple quantitative fixing [Fixierung] of a physical fact draws it into a network of theoretical presuppositions, outside of which the very question as the measurability of the process could not be raised" (SF 142–143/155, emphasis in original). More specifically, Cassirer argues that instead of simply being given facts of measurement, natural science must arrive at them through a theory-dependent process in which it transforms what is empirically present into an object that is measurable, that is, a physical object in the precise sense. As he makes this point, what is empirically present must be "subjected to a transformation [Umdeutung], by which it is changed into another type of manifold, which we can produce and master according to rational laws" (SF 116/125, translation modified). According to Cassirer, this transformation is the result of us unifying, ordering, and organizing what is empirically present ('the empirical manifold') by means of a function-concept.

Cassirer illustrates the function-dependent nature of measurement with an analysis of the measurement of motion (see SF 117–122/126–132). In order to measure the motion of a body, Cassirer indicates that the body must be "definitely determined, distinguished and limited from all other structures [Gebilde]" (SF 120/129). However, he asserts that such a body is one that perception never presents us with:

> the bodies of our world of perception never satisfy this condition. They owe their determinateness merely to a first and superficial unification [Zusammenfassung], wherewith we unite [vereinen] into a whole parts of space that seem to possess approximately the same sensuous properties.
>
> (SF 120/129)

If we are to be in a position to measure the motion of a physical body, Cassirer argues that we must use geometrical concepts to transform what is sensuously present to us into a "'rigid' body of pure geometry" (SF 120/130). Following Karl Pearson, he maintains that this involves a "process of idealization [*Idealisierung*]" in which we apply geometrical concepts to manifolds that are empirically present and so transform them into "*ideal limiting structures* [*Grenzgebilden*]" (SF 127/138, 121/130, emphasis in original). And, according to Cassirer, it is this ideal limiting structure that we can predicate motion of and measure. Thus, for Cassirer, natural science can establish facts about the measurement of motion only on the basis of the concept-guided transformation of what is empirically present into an ideal limiting structure. And he takes this to be an illustration of the broader truth that establishing facts of measurement, in any case, depends on relying on function-concepts to transform what is empirically present into something measurable.

From Cassirer's point of view, if even the most basic facts of measurement are dependent on function-concepts, then the substance-based approach to natural science should be rejected:

> We do not have physical concepts and physical facts [*Tatsachen*] in pure separation, so that we could select a member of the first sphere and inquire whether it possessed a copy [*Abbild*] in the second; but we possess the "facts" only by virtue of the *totality* [*Gesamtheit*] of concepts.
>
> (SF 147/159, emphasis in original)

Rather than treating facts as something brutely given, which we then attempt to copy, Cassirer endorses Goethe's view that, "All that is factual is already theory" (SF 243/262). More specifically, he argues that facts are the result of an active process in which we deploy concepts to transform what is empirically present into mathematically determinable objects. Hence Cassirer's assertion that,

> it is precisely that sort of transformation [*Veränderung*] of the initial existence [*Bestandes*], which constitutes the character and value of the intellectual procedure of physics. Instead of a mere *passive* reproduction, we see before us an *active* process, which transports what is at first given into a new logical sphere.
>
> (SF 122/, my emphasis, translation modified)

In order to give an adequate account of the actual practice of natural science, Cassirer thus concludes that we must embrace a function-based view, which acknowledges the role that function-concepts play in the active process of transformation.

What is more, once we appreciate that even the elementary facts of measurement depend on functions, Cassirer argues that the parallels between mathematics and natural science become more apparent. Indeed, he claims that the foregoing considerations reveal that, like the concepts of mathematics, the concepts of natural science should be understood as function-concepts on the basis of which ideal objects are generated. To be sure, Cassirer acknowledges that whereas in the mathematical case the logical structures that are generated are purely ideal, in the natural scientific case the ideal limiting structures are generated on the basis of what is empirically present. Nevertheless, insofar as the concepts of both mathematics and natural science rest on functions or relations that generate ideal structures, he thinks that there is continuity between the two. And in his philosophy of natural science, he thus endeavors to extend the function-based line of thinking he defended with respect to facts of mathematics to the facts of physics and chemistry.

Cassirer's critical approach to natural science

Let's turn now to the details of Cassirer's functional account of natural science. In this section, I consider how Cassirer deploys critical idealism and the critical theory of cognition in this context and in the following two sections I take up his account of the method and progress of natural science.

While I addressed how critical idealism and the critical theory of cognition shape Cassirer's account of natural science in broad strokes in Chapter Two, I shall now explore how he fills out the details. To begin, as we just saw, Cassirer takes on the basic commitment of critical idealism insofar as he maintains that the objects of natural science are made possible by functions. Returning to a passage I quoted in Chapter Two, Cassirer claims,

> we determine the object not as an absolute substance beyond all cognition, but as the object shaped in progressing experience . . .

it remains strictly within the sphere, which these principles [of logic] determine and limit, especially the universal *principles* of mathematical and scientific cognition. This simple thought alone constitutes the kernel of critical "idealism."

(SF 297/321, translation modified, emphasis in original)

Here, Cassirer begins by rejecting a substance-based view of the objects of natural science as 'absolute substances beyond all cognition'. He, instead, argues that these objects are 'relative' to the functions of 'progressing experience' (SF 298/321). Recall that, in this context, Cassirer defines 'experience' in Marburg terms, that is, not in terms of our everyday experience of the world, but rather in terms of the experience involved in natural science as a whole (see Chapter Two). So understood, the experience of natural science transcends any single individual and encompasses the body of cognition that continuously unfolds as natural science develops.[4] Thus, when Cassirer asserts that the objects of natural science depend on the functions of 'progressing experience', he means that they depend on the functions operative in mathematical-scientific cognition as a whole.

Cassirer summarizes his critical idealism with respect to the objects of natural science in the following passage from *Substance and Function* (which he quotes again in *Determinism and Indeterminism*):

we do not cognize [*erkennen*] "objects" as if they were already independently determined and given *as objects*,—but we cognize [*erkennen*] *objectively*, by producing certain limitations and by fixating certain permanent elements and connections within the uniform flow of experience. . . . The "thing" is thus no longer something unknown, lying before us as a bare material, but is an expression of the form and manner of conceiving [*Begreifens*].

(SF 303/328, translation modified, emphasis in original; quoted again in DI 137/166)

Here, we find Cassirer rejecting a substance-based account of the objects of natural science, in favor of a function-based account, according to which these objects conform to the functions of mathematical-scientific experience. And, as we saw above, he takes this to be the case because he thinks those functions are required in order to transform what is empirically present to us into a mathematically-determinable physical object.

Cassirer, in turn, couples this critical idealism with a critical theory of mathematical-scientific cognition, which aims at elucidating the functions of scientific experience. One of Cassirer's preferred ways to describe these functions of scientific experience is as 'invariant relations'. The notion of an invariant relation should sound familiar from Chapter Three and Cassirer's account of geometry. And the reasons Cassirer deploys this notion again in his account of natural science is because, on his view, just as the geometer seeks the relations that remain 'invariant' through projection and transformation, so too does the natural scientist seek relations that remain 'invariant' across all physical phenomena and scientific experience. Hence Cassirer's claim that, "The goal of all empirical cognition lies in gaining ultimate invariants" (SF 273/294, translation modified).

Cassirer, moreover, maintains that the 'cop' or 'objectivity' of our cognition in natural science needs to be understood in terms of these invariant relations or functions (see SF Ch. VI, ETR Ch. 3, PSFv3 500/552/475–476, AH 68–74). To this end, he argues that cognition in natural science is 'true' or 'objective' to the extent that it represents the relations or functions that remain invariant across physical phenomena and scientific experience. As we saw him make this point above, "we cognize [erkennen] objectively, by producing certain limitations and by fixating certain permanent elements and connections within the uniform flow of experience" (SF 303/328). Thus, rather than measuring the objectivity or truth of natural science in terms of the 'correspondence' between our theories and an absolutely mind-independent reality, Cassirer treats invariance across all physical phenomena and scientific experience as the standard. So understood, one theory will be more objective than another if it can account for a wider domain of phenomena and scientific experience.

This said, on Cassirer's view, we cannot expect to develop a mathematical-scientific theory that, in fact, captures all the invariant relations that govern all physical phenomena and all scientific experience. Instead, he regards such a theory as a 'regulative' ideal that we strive towards, and he claims that we measure the truth or objectivity of actual scientific theories in terms of how closely they approximate this ideal. Theories that represent relations that remain invariant across a narrow sphere of physical phenomena and scientific experience will be less objective and true than those that remain invariant across a broader sphere. And, as we shall see below, on Cassirer's view, progress in cognition in natural science is to be understood in terms of an advance toward theories that are truer and more objective.

In these ways, Cassirer's critical idealist commitment to the objects of natural science being dependent on the functions or relations of scientific experience, on the one hand, and his critical epistemological commitment to elucidating those functions or relations, on the other, lay the groundwork for his own functional theory of natural science. And it is within this critical framework that Cassirer develops an account of the method natural science calls for and the progress it involves.

Method and natural science

In order to understand Cassirer's approach of the methodological issues surrounding natural science, it is important to situate his account against the Neo-Kantian backdrop I discussed in Chapter Two. Recall that one of the main motivations for going 'back to Kant' was a concern over the ongoing value of philosophy in the face of natural science. And the Neo-Kantians embraced Kant's method as a way to avoid speculative metaphysics by remaining grounded in concrete facts, but without falling into the subservient position of positivism. Rather than acting as the handmaiden of natural science, the Neo-Kantians endorsed a methodology, according to which philosophy has an autonomous task to perform in relation to natural science, viz., the task of offering an account of the conditions of the possibility of cognition in natural science.

This background is salient because as we now turn to Cassirer's remarks on the methodological issues pertaining to natural science, it is clear that he not only wants to offer an account of the method of natural science itself, but also of the methodological role philosophers have to play in relation to natural science. Indeed, as we shall see, Cassirer conceives of the natural scientist and transcendental philosopher as engaged in a shared project: they seek to elucidate the invariant relations of physical phenomena and scientific experience. And he claims that each makes a distinctive contribution to this end: the natural scientist uncovers invariant relations of the physical variety, while the transcendental philosopher sheds light on those of an *a priori* variety. As Cassirer frames this point in *Determinism and Indeterminism*, each is oriented toward making particular kinds of 'statements' [*Aussagen*]: the scientist makes 'physical' statements, which articulate physical invariant relations, and the transcendental philosopher makes 'transcendental' statements, which articulate the *a priori* invariant relations, which remain constant across all scientific experience (see DI 30/38). In order to clarify Cassirer's methodological picture, I thus begin with his account of the 'physical statements' made by scientists, before turning to his analysis of the 'transcendental statements' made by philosophers. And at the end of this section, I consider how Cassirer's methodological account of these various 'invariant relations' sheds light on the debate over his conception of the *a priori* involved in natural science.

Physical statements

In order to clarify the 'immanent methodological coherence' of natural science Cassirer organizes his discussion around an analysis of three different types of 'statements' that scientists make in which they attempt to express the physical invariant relations that govern physical phenomena. He refers to these as "statements of measurement" (*Massaussagen*), "statements of laws" (*Gesetzesaussagen*), and "statements of principles" (*Prinzipienaussagen*) (DI 30/38). As we shall see below, he aligns each kind of statement with a different kind of physical invariant relation and type of physical phenomena: he connects statements of measurement to physical concepts and 'individual' physical phenomena, statements of laws to physical laws and 'general' physical phenomena, and statements of principles to physical principles and 'universal' physical phenomena. And although

Cassirer discusses each type of statement individually, he argues that these statements reciprocally support one another: "all the statements of physics are determined through one another, they mutually condition and support one another. . . . This reciprocal interweaving and bonding constitutes one of the basic features of the system of physics" (DI 35/44). Cassirer thus conceives of the method of natural science in terms of this overarching effort to make physical statements of all three sorts, which reflect the physical concepts, laws, and principles that govern physical phenomena. And he insists that this holistic picture should be borne in mind when considering each particular type of statement.

Statements of measurement

As we have already seen, Cassirer thinks that a methodological commitment to measurement serves as the foundation of mathematical natural science. He thus accords 'statements of measurement' a privileged place. As he makes this point in relation to physics, "the statements of measurement may indeed be designated as the alpha and omega of physics, its beginning and end. From them all its judgments take their departure and to them they must all lead back again" (DI 36/45, translation modified). However, on Cassirer's view, statements of measurement serve as the 'alpha and omega' not just of physics, but of all mathematical natural science, as it is oriented from first to last around the mathematical determination of physical phenomena through measurement.

According to Cassirer, what is distinctive about statements of measurement is their *individuality*: "What characterizes statements of the first level and distinguishes them sharply from all others is the feature of *individuality*; they all refer to a definite point of space and time to which, so to speak, they are bound" (DI 36/46, emphasis in original). A statement like, 'The gas in this tube is 20°C', is an example of a statement of measurement insofar as it assigns a numerical value to an individual, which occupies a definite point in space and time.

Yet, as we saw above, on Cassirer's view, a statement of measurement does not simply passively register the given facts; it is the result of an active intellectual process in which what is empirically present is transformed into an ideal limiting structure that can be mathematically measured. And, according to Cassirer, there is a distinctive set of invariant relations that natural science relies on to carry out this transformative work, viz., physical 'concepts':

> concepts, such as those of mass and force, the atom or the ether, the magnetic or electrical potential, even concepts, like those of pressure or of temperature, are no simple thing-concepts, no copies of particular contents given in perception. . . . What we possess in them are obviously not reproductions of simple things or sensations, but the theoretical assumptions and constructions, which are intended to transform the merely sensible into something measurable, and thus into an "object of physics," that is, into an object for physics.
>
> (ETR 357/7–8)

Here, Cassirer maintains that it is by applying a concept, like 'mass' or 'temperature', to what is empirically present that natural science transforms it into something measurable, hence into a proper physical object. On Cassirer's view, it is thus only on the basis of transformation via physical concepts that statements of measurement become possible.

Consider, for example, Cassirer's Duhem-inspired analysis of the concept of temperature (see SF 142–143/153–155). According to the physical concept of temperature, there is an exact correlation between heat and extension, and thermometers reflect this theoretical assumption. When a scientist uses a mercury thermometer, for example, the reason that she can read off a temperature from a height of mercury is because she is guided by the physical concept of temperature, hence the theoretical presupposition that different volumes of mercury are directly proportional to different Celsius degrees. And it is on account of this concept that she can treat the height of mercury that is empirically present to her as having a temperature value. Applying this line of thinking to statements of measurement, it is only by transforming the empirically present manifold in light of the concept of temperature that a scientist can arrive at a statement of measurement, like 'The gas in the tube is 20°C'.

For Cassirer, then, the effort to make statements of measurement is not just a matter of measuring what is empirically present. It requires developing the physical concepts, which are a pre-requisite of a scientist being able to transform what is empirically present into something that is measurable.

Statements of laws

Unlike the individual orientation of statements of measurement, Cassirer argues that statements of law are oriented in a more general direction. With them, the scientist seeks to articulate invariant relations that govern entire 'classes' of physical phenomena, such as, the class of mechanical phenomena or of electro-dynamical phenomena (DI 44/55). In this vein, he says,

> The mere "Here-thus" that is contained in the particular statements of measurements undergoes a characteristic transformation in the statements of laws: it is converted into an "If-then." And this "If-then," this hypothetical judgment, If x, then y, does not merely combine particular quantities which we consider as belonging to and localized in definite points in space-time, but pertains rather to whole classes of magnitudes, classes which consist generally of infinitely many elements.
>
> (DI 41/51, translation modified)

Cassirer, in turn, connects statements of laws to the practice of experimentation. He argues that the aim of an experiment is to uncover the lawful connections that govern a class of phenomena (see DI Ch. 4, SF Ch. 4). Thus, in the experiment, what is of primary interest is not the individuals present here and now for their own sake, but rather the law that obtains for the phenomena belonging to a particular class, which those individuals are instances of. Accordingly, Cassirer claims that scientists are interested in the results of the experiment as something that can be "transferred from place to place, instant to instant" (DI 41/52).

Once again, however, Cassirer emphasizes that the practice of experimentation is not free from theoretical presuppositions. Though physical concepts continue to play a role, Cassirer argues that experiments presuppose another kind of invariant relation, viz., a physical law, which the scientist presupposes as a hypothesis and then looks to the observed phenomena to confirm. He claims that this anticipated law plays a two-fold role in experimental observations. On the one hand, it guides scientists toward which measurements to make. On the other hand, it shapes the overall perspective of the scientist, putting her in the position to see what is present here and now as a 'symbol' of an entire class of phenomena and, accordingly, to transfer the results from the here and now to other phenomena that are not present (SF 247/266).

As an example, let's take Henri Regnault's experiment to test the Boyle-Mariotte law (PV = K), according to which if there is a constant temperature, then the pressure and volume of a gas will vary proportionally (see SF 143–144/155–156). According to Cassirer, in his experimentation, Regnault presupposes this law as a hypothesis and it directs him towards taking the relevant measurements of temperature, pressure, and volume. Moreover, insofar as his outlook is shaped by this law, Regnault is able to treat the results of any particular experiment as symbolic of, hence transferrable to, the class of gases as a whole.

Far from simply recording observations, then, Cassirer suggests that experiments of this sort involve a great deal of intellectual work with respect not just to the initial formulation of those laws, but also to how those laws guide measurement practices and the symbolic interpretation of experimental results. In these manifold ways Cassirer takes the invariant relation of a physical law to shape the practice of experimentation and the scientific endeavor of issuing statements of laws.

Statements of principles

With statements of principles, Cassirer claims that the scientist engages in an endeavor that is universal in orientation, as she seeks to uncover the invariant relations that govern all physical phenomena, regardless of which class they belong to. As he puts it, these statements are oriented by "certain common denominators valid for all phenomena" (DI 52/65). So understood, statements of principles apply to nature as a whole.

However, according to Cassirer, statements of principles do not directly refer to phenomena, but rather to the functional connections between physical laws. As he makes this point in Substance and Function, each physical law (ψ) represents a functional relation that governs a particular class of physical phenomena, and what each physical principle (φ) does is connect those functional relations together by means of a "more inclusive" or universal relation, for example, $\varphi(\psi_1(\alpha_1, \alpha_2 \ldots \alpha_n), \psi_2(\beta_1, \beta_2 \ldots \beta_n))$ (SF 267/288). Maxwell's equations, for example, count as physical principles insofar as they assert a functional connection between the physical laws that govern magnetism, electricity, and light (see DI 45/57, PSFv3 511–512/506–507/437–438).

Although physical principles thus assert a connection between different physical laws, Cassirer argues that they initially do so only in a hypothetical way:

> The principles of physics are basically nothing but such means of orientation, means for surveying and gaining perspective. At first they are only valid as hypotheses. They cannot stimulate dogmatically from the beginning a particular result of investigation. But they teach us how to find direction in which we have to advance.
>
> (DI 53–54/67)

For this reason, Cassirer suggests that statements of principles are best understood from a "heuristic point of view," as anticipating, but not guaranteeing, a unified picture of different domains of physical phenomena (DI 52/65).

Consider, for example, the principle of least action. Cassirer credits Helmholtz for recognizing the heuristic value of this principle as one that promises to shed light not just on mechanical phenomena, but on all physical phenomena: "the universal validity of the principle appears to me to be so far assured that it may claim considerable importance as a working hypothesis and as a guideline in the effort to formulate laws for new classes of phenomena" (DI 49/62, quoting Helmholtz's "On the Physical Significance of the Principle of Least Action"). Highlighting the effectiveness of this principle as a guideline in modern physics, Cassirer points out that Planck references it as "the highest *physical* law which contains the four world coordinates in a completely systematical arrangement" (DI 49/62, emphasis in original). And he notes that this principle paves the way for the development of the "'generalized law of inertia'" in general relativity and for Schrödinger's formulation of the analogy between geometrical and wave optics in quantum mechanics (DI 50/63). For Cassirer, then, the principle of least action is a physical principle that plays a heuristic role in the method of natural science, guiding it toward systematic account of all physical phenomena. And, according to Cassirer, statements of principles of this sort play a crucial role in the scientific attempt to give an account of all physical phenomena.

Stepping back, on Cassirer's view, the method of natural science involves the effort to make three different kinds of physical statements, which reflect different physical invariant relations that govern physical phenomena. Statements of measurement reflect physical concepts that order individual phenomena, statements of law reflect physical laws that order general classes of phenomena, and statements of principles reflect the physical principles that functionally correlate different laws together and so organize physical phenomena in a universal way.

However, according to Cassirer, part of the methodology of natural science involves recognizing that physical statements and the invariant relations they reflect are open to revision:

> we cannot claim "absoluteness" or ultimacy for any empirical judgment however high in the order of empirical cognition it may stand, but must always leave open the possibility that an advance in cognition may lead us to supplement or correct that judgment.
>
> (DI 54/68, translation modified)

Seen from this perspective, Cassirer maintains that the practice of natural science is:

> a never-ending process through which the relatively "necessary" takes the place of the relatively accidental and the relatively invariable take the place of the relatively variable. We can never claim that this process has attained to the ultimate "invariants

of experience". . . . Rather, the possibility must always be held
open so that a new synthesis will instate itself and so that the
universal constants, in terms of which we understand large
realms of physical objects, will come closer together and prove
themselves to be the special cases of an overarching lawfulness.
This forms, then, the true core of objectivity.

(PSFv3 550/552/476)

Here, Cassirer indicates that scientists cannot expect to arrive at
physical statements that perfectly capture the physical invariant
relations that govern all physical phenomena and scientific experience. Instead, they only arrive at "relative stopping points," which
capture the physical invariant relations that govern increasingly
wider spheres of physical phenomena (SF 278/299). However, even
though these are only 'relative stopping points', Cassirer claims that
the cognition that scientists achieve at each stage is 'objective' to the
extent that the physical invariants that it articulates, indeed, govern a
sphere of physical phenomena.

Transcendental statements

On Cassirer's view, the physical invariant relations uncovered through
scientific investigation do not exhaust all the invariant relations that
bear on scientific cognition; rather, these scientific efforts need
to be supplemented by the efforts of transcendental philosophers
who alone are able to provide insight into the *a priori* invariant
relations that make scientific cognition possible in the first place.
For this reason, Cassirer claims that in order to do justice to the full
sweep of invariant relations, we need both the physical statements
of scientists and the transcendental statements of philosophers. As
he draws this contrast, "Instead of being a direct statement about
things, [a transcendental statement] must be viewed as a statement
about our empirical cognition of things,—that is, about experience"
(DI 58/72, translation modified).[5] And, as we shall see, Cassirer
treats transcendental statements as statements about the 'universal
invariants' that make empirical cognition in natural science possible.

In his analysis of this transcendental contribution, Cassirer
distinguishes between two kinds of universal invariants that can be

reflected in a transcendental statement: constitutive concepts or categories, on the one hand, and regulative principles, on the other. He thus organizes his analysis of the method the transcendental philosopher deploys with respect to natural science as an attempt to uncover these constitutive and regulative invariant relations that make scientific experience possible. In order to explore his view of this transcendental project, I start with a discussion of his view of the universal invariants of scientific experience, in general, before turning to his account of the constitutive and regulative invariant relations.

The universal invariants of experience

In *Substance and Function*, Cassirer argues that in relation to natural science, the transcendental philosopher has the task of offering a 'universal invariant theory of experience':

> the critical theory of experience would constitute the *universal invariant theory of experience*. . . . The procedure of the "transcendental philosophy" can be directly compared at this point with that of geometry. Just as the geometrician selects for investigation those relations of a definite figure [*Figur*], which remain unchanged by certain transformations, so here the attempt is made to discover those universal elements of form, that persist through all change in the particular material content of experience. . . . The "categories" of space and time, of magnitude [*Größe*] and the functional dependency of magnitudes, *etc.*, are established as such elements of form, which cannot be lacking in any empirical judgment or system of judgments. . . . The goal of critical analysis would be reached, if we succeeded in isolating in this way the ultimate common element of all possible forms of scientific experience; i.e., if we succeeded in conceptually defining those moments, which persist in the advance from theory to theory because they are the conditions of any theory.
> (SF 268–269/289–290)

Here, Cassirer claims that the transcendental philosopher should seek to offer a 'universal invariant theory of experience'. And he cashes out this project by means of a comparison with geometry (see Chapter Three): just as the geometer seeks to elucidate the invariant relations that remain constant across transformations, the transcendental philosopher seeks to uncover the invariant relations that remain constant across all scientific experience. Cassirer illustrates the notion of a 'universal invariant of experience' with the concepts of 'space', 'time', 'magnitude' (Größe), and 'causality' (SF 269/290). According to Cassirer, these concepts remain the same across all scientific experience because they are *a priori* conditions of the possibility of that experience. As he makes this point in *Einstein's Theory of Relativity*, universal invariants represent the "pure functions of form and order . . . [and] formal principles in which is expressed the universal conditions of the 'possibility of experience'" (ETR 420/82). Or, as he makes this point in *Substance and Function*,

> Only those ultimate *logical invariants* can be called *a priori*, which lie at the basis of any determination of a connection [*Zusammenhänge*] according to natural law. A cognition is called *a priori* not in any sense as if it were *prior* to experience, but because and in so far as it is contained as a necessary premise in every valid judgment concerning facts.
>
> (SF 269/290, emphasis in original)

Cassirer thus ascribes to the transcendental philosopher the task of establishing the universal, *a priori* invariants that make scientific experience possible in the first place.

This said, Cassirer also insists that no one critical philosopher can hope to offer an exhaustive account of all universal invariants: "At no stage of knowledge [*Wissens*] can this goal be perfectly achieved; nevertheless it remains as a *demand*, and prescribes a fixed direction of the continuous unfolding and evolution of the systems of experience" (SF 269/290, translation modified, emphasis in original). In this passage, Cassirer mounts an implicit criticism of Kant of the sort I discussed in Chapter Two, according to which Kant is mistaken to treat the *a priori* conditions of experience as wholly fixed and static. Instead, Cassirer advocates, in a Hegelian-style, for an open-ended account of these *a priori* invariants, which allows for them to develop historically.

Though this gives us a sense of Cassirer's picture of the sort of universal invariants at issue in transcendental statements, as I mentioned earlier, he distinguishes between two types of universal invariants: constitutive concepts ('categories') and regulative principles. The constitutive-regulative distinction has its origins in Kant's first *Critique*, where Kant uses the term 'constitutive' to refer to concepts and principles that condition objects of cognition and 'regulative' to refer to concepts and principles that merely 'regulate' our empirical inquiry (see Chapter Two for a discussion of regulative principles). As Cassirer describes this distinction in *Kant's Life and Thought*, constitutive representations are ones that "determine and anticipate what is given in the object," and regulative representations are ones that provide the "rule of progress" for scientific experience and a "guideline for empirical inquiry" (KLT 205–207/198–200). On Cassirer's interpretation, then, the constitutive-regulative distinction is a distinction between the representations that are constitutive for objects of cognition because they serve as the conditions of the possibility of any particular scientific experience, and representations that are regulative for the progress of experience as a whole. Cassirer, in turn, utilizes this distinction in his account of transcendental statements, distinguishing between those that reflect constitutive concepts, which I shall just refer to as 'categories', and regulative principles.

The categories of experience

According to Cassirer, in her endeavors to elucidate the categories of experience, the transcendental philosopher seeks to uncover the concepts that are constitutive of the objects of scientific experience, that is, of mathematically-determinate physical objects. The categories 'determine' and 'anticipate' these objects insofar as they serve as the basic relations natural science relies on in transforming what is empirically present into a mathematically-determinate physical object. As Cassirer makes this point in *Substance and Function*, the categories are "invariant[s] in every *physical construction*" (SF 270/291, emphasis in original).

We have already encountered some of Cassirer's candidates for categories in his discussion of the universal invariant theory of experience, where he lists space, time, magnitude, and the functional dependency of magnitudes, that is, causality, as the 'categories' requisite for every physical construction (SF 269/290).[6] And though he does not mention 'number' on this list in *Substance and Function*, his frequent mention of it elsewhere indicates that he takes it to be another category of scientific experience (see, e.g., SF 309/334; ETR 420/81, 445/111). The categories of space, time, magnitudes, causality, and number thus serve as examples of concepts Cassirer takes to be constitutive of the objects of scientific experience.

This being said, Cassirer distinguishes between two aspects of the categories of experience, which I shall call their 'logical form' and 'content'. On his view, the logical form of a category refers to the basic logical relation that it involves. For example, the logical form of space is the relation of 'coexistence' (*Nebeneinander*), time is the relation of 'succession' (*Nacheinander*), and causality is the relation of "space-time dependency of the elements in natural processes" (ETR 418/79, see also ETR Chs. V–VI). And according to Cassirer the logical form of the category is one that remains constant across all scientific experience, hence across all theory change.

By contrast, he takes the particular content of a category to reflect how that category is articulated within a particular theory. For example, the spatial relation of coexistence can take on Euclidean or Minkowskian content. Or the logical form of causality takes on content through "*particular* causal principles" (SF 269/290, emphasis in original).

Appreciating the distinction between the logical form and content of the categories is crucial for understanding Cassirer's position on the stability and dynamism of the categories. On Cassirer's view, insofar as the logical form of the categories remains fixed, there is something about them that remains stable across all theory change. However, insofar as the content categories can vary from theory to theory, there is a dynamic aspect to them as well. As he makes this point in a letter to Moritz Schlick in 1920, the *a priori* is "a function that . . . remains the same regarding its *direction* and its *form*; nevertheless, it can assume the most various developments in the progress of cognition" (Corr. 50–51, my translation). Cassirer's account of the categories thus blends together a Kantian-style insight that the categories have a stable logical form and a Hegelian-style insight that the content of the categories is something that dynamically develops. And on his view, one of the tasks of the transcendental philosopher is to make transcendental statements, which reflect the categories in all their complexity.

Regulative principles of experience

In addition to making transcendental statements that express the categories, Cassirer claims that the transcendental philosopher should also endeavor to make statements that reflect *a priori* regulative principles that guide scientific experience as a whole. Cassirer illustrates the sorts of transcendental statements he has in mind with a discussion of three principles: the principles of systematicity, simplicity, and causality.

According to Cassirer, the principle of systematicity articulates the demand that scientific experience as a whole progress towards theories that are continuous and unified (see, e.g., SF 319/345). A continuous theory, on Cassirer's view, is one that has no gaps: it is able to seamlessly make sense of all individual physical phenomena, classes of phenomena, and their systematic relations to one another in nature as a whole. And a unified theory is one that can account for all scientific experience and thereby all physical phenomena. Quoting Planck to this effect, Cassirer claims that a unified theory is one that has "unity in respect of all features of the picture, unity in respect of all places and times, unity with regard to all investigators, all nations, all civilizations" (SF 307/332). Cassirer glosses this demand for unity in both a negative and a positive vein. In the negative vein, this amounts to the demand to develop theories that avoid 'anthropomorphism', that is, privileging a particular standpoint or perspective (see SF 306/331; PSFv3 504/499/431). And in a positive vein, he glosses it in terms of the demand to develop theories that are 'objective' or 'true' in the sense that they articulate the invariant relations that apply to all phenomena and all scientific experience. Cassirer thus treats the principle of systematicity as one that reflects the *a priori* demand for continuous and unified theories, which 'regulates' scientific experience as it progresses closer toward 'objectivity'.

Cassirer also considers the principle of simplicity as the subject matter of a transcendental statement, insofar as it reflects the regulative demand that natural science progress towards theories in which "the greatest possible number of phenomena is comprehended with the fewest possible determining factors" (DI 70/88). On Cassirer's view, the simplicity that is demanded is not just simplicity at the level of individual phenomena or a class of phenomena; rather it "pertain[s] to the *total system*" (DI 69/86, emphasis in original). That is to say, this principle sets as a goal for natural science a simple theory of nature as a whole, which articulates the physical concepts, laws, and principles that organize nature in the simplest fashion possible.

The final *a priori* regulative principle that Cassirer discusses at length is the principle of causality. This comes as something of a surprise since, as we just saw, in *Substance and Function*, Cassirer treats causality as a category. However, one of Cassirer's key claims in *Determinism and Indeterminism* is that causality is also a regulative principle.

Cassirer pursues the topic of the principle of causality at length in *Determinism and Indeterminism* in an effort to address certain skeptical concerns that arise with regard to the continued viability of the notion of causality within quantum mechanics. As Cassirer points out, Heisenberg himself declares that, "'the invalidity of the causal law is definitely established by quantum mechanics'" (DI 123/148, quoting Heisenberg). However, Cassirer argues that in order to adjudicate this issue, we need to distinguish between two senses of causality: causality as a metaphysical concept and as a regulative principle. As a metaphysical concept, causality refers to the existence of an entirely determinate causal order among physical phenomena, of the sort that a Laplacean spirit would be able to comprehend (see DI Chs. 1–2). By contrast, as a regulative principle, Cassirer claims that causality refers to a certain methodological demand placed on scientific experience as a whole to progress as far as possible in issuing statements of measurement, laws, and principles:

> the causal principle signifies . . . a new insight . . . concerning method. . . . [I]t is a statement "*about*" measurements, laws, and principles. It says that all these can be so related and combined [*verknüpfen*] with one another that from this combination

[*Verknüpfung*] there results a *system of physical cognition* and not a mere
aggregate of isolated observations.
(DI 60/74–75, translation modified, emphasis in original)

So understood, Cassirer treats the regulative principle of causality
as a principle that demands the development of theories that com-
bine together physical statements of concepts, laws, and principles
in a systematic way. He often makes this point in Helmholtz's terms,
describing the systematic connection between measurements, physical laws, and physical principles in terms of maximal "orderliness
according to law in appearances [*das Gesetzliche in den Erscheinungen*]"
(DI 62/77, translation modified). And his idea is that the regulative principle of causality reflects the demand to develop theories
which reflect the overarching 'orderliness according to law', which
is expressed through all three physical statements, that govern the
physical world.

Drawing on the distinction between causality in a metaphysical and regulative sense, Cassirer argues that although quantum
mechanics undermines the metaphysical conception of causality,
it nevertheless remains wedded to the regulative principle of causality. He takes this to be the case because he thinks that physicists
operating within this framework are still responsive to the methodological demand to find as much 'orderliness according to law'
as possible at the level of statements of measurements, laws, and
principles. For this reason, Cassirer claims,

> none of these "revolutionaries" [in quantum mechanics] wanted
> to dispense altogether in their actual physical procedure with
> conformity to law [*Gesetzlichkeit*] of events *in general*; rather they
> ask how they may *express* and establish this conformity unobjec-
> tionably under the conditions of our observation of nature.
> (DI 115/138, translation modified, emphasis in original)

On Cassirer's reading, then, while quantum mechanics may undermine causality as a metaphysical concept, it does not undermine
it as a regulative principle.

Cassirer's treatment of causality as a regulative principle in
Determinism and Indeterminism raises the question of whether he thereby

revises his earlier account of causality as a category in *Substance and Function*. However, a closer look at his treatment of causality reveals that he thinks of it as both a category and a regulative principle even within the framework of quantum mechanics. As I noted above, in *Substance and Function*, Cassirer glosses causality in terms of the "functional dependency of magnitudes" (SF 269/290). And he continues to describe causality in terms of 'functional dependency' in *Determinism and Indeterminism*:

> The essence of the causal concept remains untouched as long as this essence is grasped in its true universality—that is, defined only by the demand for strict *functional dependence*. If the individual elements of determination available to quantum mechanics are used in accordance with the general principles of the theory and in keeping with the limits fixed by the uncertainty relations, a functional relation, precisely definable will always exist between them.
>
> (DI 188/225, my emphasis)

Moreover, he argues that when we understand causality in terms of function dependency, then we should recognize that there is a "causal law of quantum mechanics" according to which, "if at any time certain physical magnitudes [*Größen*] are measured as exactly as possible in principle, magnitudes [*Größen*] will also exist at any other time whose magnitude on being measured can be predicted with precision" (DI 188/225–226, translation modified). In this context, Cassirer appears to have in mind the Born rule, which allows for the calculation of the probabilities of particular outcomes for measurement of a given observable for a given quantum state. And the way Cassirer describes the Born rule suggests that it articulates an, albeit probabilistic, functional relationship between magnitudes. Seen from this perspective, the Born rule looks less like a regulative principle for the progress of natural science as a whole and more like a category, which articulates the functional dependency of magnitudes in the quantum framework. In *Determinism and Indeterminism* Cassirer thus discusses two different notions of causality, causality as a category and causality as a regulative principle.

Caveats aside, as a regulative principle, Cassirer treats causality as an *a priori* regulative principle that directs the progress of natural science as a whole towards developing theories that articulate the maximal 'orderliness according to law' of physical phenomena and systematically elucidate the invariant relations that govern nature through physical statements of measurements, laws, and principles. And he thus treats the principle of causality as a regulative principle, alongside the principles of systematicity and simplicity, which serves as the subject matter of transcendental statements about the regulative principles that guide scientific experience as such.

Stepping back, on Cassirer's view, the transcendental philosopher has a certain methodological task in relation to scientific experience: she is to articulate the *a priori* invariant relations, the categories and regulative principles, that serve as conditions of the possibility of scientific experience as such. By Cassirer's lights, these philosophical efforts serve as a complement to the efforts made by natural scientists to uncover the physical invariant relations, the physical concepts, laws, and principles, which govern physical phenomena. And, according to Cassirer, it is only by means of both transcendental and physical statements that we can hope to fully understand the invariant relations that underwrite scientific cognition.

Cassirer and the variety of the a priori

Before turning to Cassirer's account of scientific progress, I want to pause and consider the question of how exactly he conceives of the *a priori* elements of natural science. This is one of the most controversial features of Cassirer's view, as commentators debate what kind of *a priori* Cassirer endorses (see Chapter Eight). The disagreements arise on two fronts. First, there is a debate over whether Cassirer conceives of the *a priori* elements of natural science in constitutive or regulative terms. This contrast is between the *a priori*, qua what is constitutive of objects of scientific experience and the *a priori*, qua what merely regulates the progress of scientific experience.[7] Second, there is a debate about whether Cassirer understands these *a priori* elements in absolute or relativized terms. Does he conceive of the *a priori* as something that is absolute, hence fixed and constant across all scientific theories, or does he, like Hans Reichenbach, conceive of the *a priori* in relativized terms as something that changes from theory to theory?[8]

When we approach the question of Cassirer's theory of the *a priori* through the lens of his methodological framework, each kind of *a priori* has a role to play. Starting with his account of transcendental statements, as we have just seen, he distinguishes between *a priori* invariants of the constitutive and regulative type, viz., the categories of experience, which are constitutive of the objects of natural science, and regulative principles, which guide scientific experience as such. He, moreover, appears to conceive of these *a priori* invariants in both absolute and relativized terms. Insofar as Cassirer regards the regulative principles as conditions of the possibility of scientific experience as such, they appear to be absolute. As for the categories of experience, since he treats their logical form as constant across theory change, this form appears to be an absolute *a priori*. However, given that the content of the categories changes from theory to theory, he seems to understand this content as a kind of relativized *a priori*.

Even at the level of Cassirer's accounts of physical statements, we can find the different notions of the *a priori* at work, albeit in a more attenuated sense. Unlike the *a priori* invariants, which are conditions of the possibility of scientific experience as such, Cassirer treats the invariant relations at issue in physical statements, viz., physical concepts, laws, and principles, as ones that vary from theory to theory. Nevertheless, insofar as Cassirer is committed to the idea that 'all that is factual is theory', he treats physical concepts, laws, and principles as a kind of *a priori*: they serve as conditions of transformation on the basis of which physical facts become possible in the first place. Cassirer thus appears to treat physical concepts, laws, and principles as a kind of *a priori* that is relativized to a particular theory. However, within this relativized framework, we saw that he treats physical concepts and laws as constitutive relations on the basis of which we construct particular types of physical objects, and physical principles as regulative relations, which guide scientific investigation in a particular direction. Thus, even within his account of physical statements, he appears to acknowledge a relativized *a priori* of both the constitutive and regulative type.

Approaching Cassirer's account of the *a priori* through this methodological lens thus points toward a role for the constitutive, regulative, absolute, and relativized *a priori* to play on his view.

The progress of natural science

The last aspect of Cassirer's account of natural science that I shall
discuss in this chapter is his account of scientific progress. Unlike
other philosophers of science, like Thomas Kuhn and Karl Popper,
who take the paradigm shifts in the 20th century to reveal the dis-
continuity of scientific progress, Cassirer defends a view of scientific
progress as something that is continuous.

In making his case for the continuity of scientific progress,
Cassirer argues that this progress is best understood in terms of a
continuous, systematically ordered series of theories. In this vein, he
claims that, "Even in the loosest, most slack succession of members,
the preceding member is not absolutely destroyed by the entry of
the succeeding member; but certain fundamental determinations
persist, on which rest the homogeneity and uniformity of the
series" (SF 266/287). Cassirer, in turn, identifies five different kinds
of 'fundamental determinations' that account for the 'homogeneity'
and 'uniformity' of this series: formal features, material features, an
interrogative structure, convergence, and a process of correction.

Cassirer glosses the formal features of the series in terms of the
"'form' of experience as a whole ['Form' der Gesamterfahrung]" that
conditions every mathematical-scientific theory (see SF 268/321,
translation modified). We can tease out Cassirer's various features of
the 'form of experience' on the basis of the preceding discussion.
Part of this 'form' is methodological in nature. At the most basic
level, the form of mathematical natural science is oriented towards
the mathematical determination of nature. This form is, moreover,
structured around the attempt to make physical statements and
uncover physical invariant relations.

In addition to this methodological form, Cassirer treats categories
and regulative principles as part of the 'form' of scientific experience.
Emphasizing this point with respect to the categories, Cassirer argues
that they provide a stable framework in which theories dynamically develop. For example, the logical form of space as a relation of

coexistence remains the same across Newtonian mechanics and relativity, and it is within this stable framework that the dynamic shift from conceiving of space in Euclidean terms to Minkowskian terms takes place.[9] So too in the case of magnitude, Cassirer maintains that all scientific experience is structured by the logical form of a numerically determinate object, and this provides the stable background against which the content of magnitude is understood in dynamically different ways in classical mechanics, relativity, and quantum mechanics, as the observer's role is increasingly taken into account. In both cases, Cassirer thinks that these dynamic shifts in the content of the categories should be situated in a stable formal structure of scientific experience provided by the logical form of the categories. And, in this way, Cassirer regards the categories as serving, alongside regulative principles, as invariants that formally structure scientific experience at the *a priori* level, and, thereby, account for part of the continuity in the progress of scientific theories.

In addition to these formal structures, Cassirer maintains that there are material features that account for the continuity of the series of scientific theories. The material features he has in mind are the empirical facts that are established through scientific experimentation and research. For example, he claims, "All the observations of Tycho de Brahe enter into the system of Kepler, although they are connected and conceived in a new way" (SF 321/348). And this responsibility to do justice to previous empirical research is something that Cassirer thinks binds the scientific theories within the series together.

The third continuity-generating feature that Cassirer highlights is the interrogative structure of the series. On Cassirer's view, newer theories are responsible for answering questions that older theories pose: "The new form [i.e., theory] must contain the *answer* to *questions*, proposed within the older form; this one feature establishes a logical connection between them, and points to a common forum of judgment, to which both are subjected" (SF 268/289, emphasis in original). Cassirer regards this interrogative structure as one of the features which unify theories together as we progress from one to another.

The next continuous feature that Cassirer calls our attention to is the convergence of the progress of scientific experience. In this vein, he argues that, as the series progresses, scientific theories converge towards an 'ideal limit':

> Knowledge [*Wissen*] realizes itself only in a succession of logical acts, in a series that must be run through successively, so that we may become aware of the rule of its progress. But if this series is to be grasped as unity, as an expression of an *identical* state of affairs [*Sachverhalts*], which is defined the more exactly the further we advance, then we must conceive the series as converging towards an ideal limit.
>
> (SF 315/341, emphasis in original, translation modified)

Part of Cassirer's conception of this ideal limit turns on a general conception of the overarching goal of mathematical science in terms of gaining insight into the mathematical determination of physical phenomena. However, Cassirer also cashes out this ideal limit in terms of the goal that guides science through the regulative principles of systematicity, simplicity, and causality. For Cassirer, then, the 'ideal limit' that the series of scientific theories is converging towards is an objective theory of physical phenomena that is maximally systematic and simple in its articulation of the physical invariant relations that govern the whole of nature.

On Cassirer's view, if we look at the historical developments in natural science, we find that that it is, indeed, progressing towards this ideal limit at different levels. To name but a few examples, in chemistry Cassirer argues that we can trace a shift from chemistry as a more classificatory science to a more systematic, mathematical science. To this end, he claims that the periodic system of elements paves the way for a systematic and deductive approach to chemistry and it does so in an increasingly accurate fashion with the use of nuclear charge instead of atomic weight as the ordering principle (see SF 216–220/235–240; PSFv3 515–518/511–514/441–443).

Meanwhile in physics, Cassirer argues that the shift away from the invariants in classical mechanics, like Newton's laws of motion and Galilean transformations, to the new invariants in special relativity, like the law of the constancy of the propagation of light, Einstein's formulation of the principle of relativity, and Lorentz transformations, paves the way for a more systematic explanation of physical phenomena in virtue of enabling us to account for them in all inertial reference frames (see ETR Ch. 2). Or consider the invariants of general relativity. According to Cassirer, they provide a systematic and simple way to account for the equivalence of laws of nature in reference frames whatever state of motion they may be in (see ETR Ch. 2). In each of these cases, Cassirer maintains that we find theories converging closer towards the ideal limit theory in virtue of uncovering new physical invariants that provide an increasingly more systematic and simple explanation of physical phenomena.

Finally, Cassirer claims that the series of scientific theories involves a process of correction that gives it continuity as well. According to Cassirer, rather than newer theories annulling older theories, as science advances we develop theories that are more general and inclusive and that can accordingly accommodate earlier theories as 'special cases', which are valid only within a narrow sphere (see, e.g., ETR 372/26, 407/67). Thus, Cassirer claims that as natural science advances, "The earlier results are not thereby rendered valueless, but are rather confirmed within a definite sphere of validity" (SF 278/299/67). And this process of correction preserves the older theories in some way in the newer ones, again, fostering continuity in the progress of science.

For example, Cassirer argues that special relativity does not invalidate Newtonian mechanics; instead, it reveals that the validity of Newtonian mechanics is restricted to reference frames at lower velocities and that the Galilean transformations that govern those reference frames are a "special case" of the Lorentz transformations (ETR 372/26). Similarly, instead of thinking that general relativity cancels out special relativity, Cassirer, following Einstein, maintains that the former explains the latter as having validity with respect to "a definitely limited field, namely, to the phenomena in an approximately constant field of gravitation" (ETR 378/33). Even in the case of quantum mechanics, Cassirer argues that even though some, like Planck, think it involves a "breach" from classical mechanics, there is, in fact, continuity through correction here as well (DI 109/131, quoting Planck). To this end, he emphasizes Bohr's complementarity principle, according to which quantum mechanics mirrors classical mechanics within the limit of large quantum numbers (DI 111–112/133–134). He also points toward Heisenberg's view that classical physics is valid "[w]ithin the measuring apparatus," to which quantum mechanics ascribes a corresponding matrix that is related to other matrices through equations of motion (DI 121/145–146). In these attempts, Cassirer claims that we see that the results from earlier theories are not nullified, but rather are recognized as valid with respect to a narrower sphere than was originally supposed. According to Cassirer, then, the process of correction contributes in a significant way to the continuity of the progress of natural science, as newer theories do not invalidate or entirely break with earlier theories, but rather index those theories to a narrower sphere of validity.

Ultimately, for Cassirer, it is on account of these formal, material, interrogative, convergent, and corrective features that we have reason to think of progress in natural science along continuous lines, as the unfolding of a unified and homogeneous series, even in its most revolutionary moments.

Summary

In this chapter, I examined Cassirer's philosophy of natural science and his effort to update the Kantian picture in order to accommodate the revolutionary developments in physics and chemistry. As was the case in his account of mathematics, Cassirer argues that in order to account for the facts of natural science, we need to embrace a function-based, rather than a substance-based framework. To clarify his function-based approach, I began with a discussion of the role critical idealism and critical theory of cognition play in his account of natural science. I then turned to an analysis of the methodological picture that Cassirer develops with respect to natural science. To this end, I highlighted the contribution that he thinks both the natural scientist and the transcendental philosopher make in the search for the 'invariant relations' of scientific experience, the former giving us insight into the invariant relations of the physical variety (physical concepts, laws, and principles), and the latter into those of the *a priori* variety (categories, regulative principles). And I argued that when we approach the question of Cassirer's theory of the *a priori* through the lens of this methodological picture, then

we find that his framework accommodates the *a priori* in the constitutive, regulative, absolute, and relativized senses. Finally, I looked
at how this functional picture of the method of natural science
shapes Cassirer's account of the progress of natural science as
something that is continuous, even across radical paradigm shifts.
And I considered his elucidation of this continuity in terms of the
series of continuous scientific theories, which are unified together
through formal, material, interrogative, convergent, and corrective
features.

In this chapter and the previous, my focus has been on Cassirer's
theory of mathematics and natural science, which served, in many
ways, as his philosophical starting point.[10] However, even in *Substance
and Function* Cassirer acknowledges the limitations of this scientific
perspective:

> The "individual" of natural science includes and exhausts nei-
> ther the individual of aesthetic consideration nor the ethical per-
> sonalities, which are the subjects of history. For all particularity
> in natural science reduces to the discovery of *definite* [*bestimmter*]
> *magnitudes* and *relations of magnitudes*, while the peculiarity and value
> of the object of artistic consideration and of ethical judgment
> lies outside its field of vision.
>
> (SF 232–233/254, emphasis in original)

Even more pressingly in *Einstein's Theory of Relativity*, Cassirer claims
that a complete critical epistemology cannot simply focus on the
theoretical cognition we achieve through mathematics and natural
science, but must also take into account the modes of 'understanding'
involved in other regions of culture, like myth, religion, art, and
language:

> over against the whole of *theoretical* scientific cognition, [there
> is] other form- and sense-giving of an independent type and
> lawfulness [*Form- und Sinngebung von selbständigem Typus und selbständiger
> Gesetzlichkeit*], such as the ethical, the aesthetic "form." It appears
> as the task of truly *universal* critique of cognition not to level
> this manifold, this wealth and variety of forms of cognition and

understanding. . . . It is the task of systematic philosophy . . . to grasp the *whole* of the symbolic forms [*Ganze der symbolischen Formen*].

(ETR 446–447/112–114, translation modified, emphasis in original)

With these remarks Cassirer anticipates the direction in which his philosophical system will develop, viz., toward a critical philosophy of culture. And though mathematics and natural science continue to play a crucial role in this philosophy of culture, Cassirer treats them as moments of a more complex cultural system, a system of 'symbolic forms'.

Notes

1   As I noted earlier, in my account of Cassirer's philosophy of natural science, I focus on his treatment of the 'mathematical' natural sciences, viz., physics and chemistry.

2   For Cassirer's historical analysis of the developments in natural science from the Renaissance through quantum mechanics, see the four volumes of *Das Erkenntnisproblem*.

3   For his discussion of measurement, see SF Chap. 4, ETR Chap. 1, and DI Chap. 3.

4   See, e.g., Cohen (1922): 62–65 and Natorp (1921): 204–205.

5   In defining a transcendental statement this way, he takes his cue from Kant's definition of 'transcendental' cognition as "cognition . . . that is occupied not so much with objects but rather with our mode of cognition of objects" (CPR A11/B25, see DI 17/26).

6   By identifying space and time as categories, Cassirer follows his Marburg Neo-Kantian predecessors in glossing space and time in an intellectualist fashion (see Chapter Two).

7   While commentators like Ryckman (2005): 46 and Heis (2014b): §2 attribute to Cassirer a view of the *a priori* that has both constitutive and regulative aspects, others like Friedman (2000): 117, 123 and Ferrari (2012) argue that he defends only a regulative picture.

8   While commentators like Richardson (1998): Ch. 5, Ryckman (2005): Ch. 2, and Ferrari (2012), (2015): 272–277 emphasize how Cassirer anticipates Reichenbach's conception of the relativized *a priori*, others, like Heis (2014b): §2, claim that his view contains both relativized and absolute elements.

9   One of Cassirer's main aims in *Einstein's Theory of Relativity* (esp. Chs. V-VI) is to show that Kant's conception of space and time as "'forms of possible experience'," is consistent with Einstein's conception of space and time as lacking

"'physical objectivity'": "when Einstein characterizes as a fundamental feature of the theory of relativity that it takes from space and time 'the last remainder of *physical objectivity*,' it is clear that the theory only accomplishes the most definite application and carrying through of the standpoint of critical idealism within empirical science itself" (ETR 412/73).

10 As I noted in Chapter One, from the outset of his career, Cassirer did not focus exclusively on topics in mathematics and natural science. In his early historical writings, including *Leibniz's System* (1902) and the first two volumes of *The Problem of Cognition* (1906/07), in addition to discussing mathematics and natural science, Cassirer addresses themes related to ethics, right (*Recht*), aesthetics, and religion.

Further reading

Ferrari, Massimo (2012). "Between Cassirer and Kuhn. Some Remarks on Friedman's Relativized A Priori." *Studies in History and Philosophy of Science* 43: 18–26. A critical discussion of Friedman's analysis of Cassirer's theory of the *a priori*.

——(2015). "Cassirer and the Philosophy of Science." In *New Approaches to Neo-Kantianism*, edited by Nicolas de Warren and Andrea Staiti, 261–284. Cambridge: Cambridge University Press. An overview of Cassirer's philosophy of science.

Friedman, Michael (2000). *A Parting of the Ways*. Chicago: Open Court. See Chapter Seven for a discussion of Cassirer's account of the *a priori* and his relation to Carnap and Schlick.

——(2005a). "Ernst Cassirer and the Philosophy of Science." In *Continental Philosophy of Science*, edited by Gary Gutting, 71–83. Oxford: Blackwell. An overview of Cassirer's philosophy of science.

——(2005b). "Ernst Cassirer and Contemporary Philosophy of Science." *Angelaki*, 10(1): 119–128. A defense of a model of the development of science progress, which takes its cue from Cassirer's conception of serial continuity and Reichenbach's relativized *a priori*, in response to Kuhn's theory of scientific revolutions.

——(2008). "Ernst Cassirer and Thomas Kuhn: The Neo-Kantian Tradition in History and Philosophy of Science." *Philosophical Forum*, Special Issue: Classical Neo-Kantianism, edited by Andrew Chignell, Terence Irwin, and Thomas Teufel, 39(2): 239–252. A discussion of the relationship between Cassirer and Kuhn on the topic of scientific progress.

Heis, Jeremy (2014). "Realism, Functions, and the A Priori: Ernst Cassirer's Philosophy of Science." *Studies in History and Philosophy of Science* 48: 10–19. A discussion of two aspects of Cassirer's philosophy of science: his account of the *a priori* and his relation to contemporary versions of structural realism (I return to these issues in Chapter Eight).

Ihmig, Karol-Nobert (1997). *Cassirers Invariantentheorie der Erfahrung und seine Rezeption des 'Erlangers Programms'*. Hamburg: Meiner. A full-length study of the relevance of Klein's Erlangen Program for Cassirer's philosophy of mathematics and natural science.

——(1999). "Ernst Cassirer and the Structural Conception of Objects in Modern Science: The Importance of the 'Erlanger Programm'." *Science in Context* 12(4): 513–529. A briefer discussion of the same topic.

Mormann, Thomas (2015). "From Mathematics to Quantum Mechanics—On the Conceptual Unity of Cassirer's Philosophy of Science (1907–1937)." In *The Philosophy of Ernst Cassirer*, edited by J. Tyler Friedman and Sebastian Luft, 31–64. Berlin: de Gruyter. An analysis of the unity of Cassirer's idealist approach to science across KMM, SF, ETR, and DI.

Richardson, Alan (1998). *Carnap's Construction of the World: The Aufbau and the Emergence of Logical Empiricism*. Cambridge: Cambridge University Press. See Chapter Five for a discussion of Cassirer's notion of the *a priori* in relation to conventionalism and Reichenbach.

Ryckman, Thomas (2005). *The Reign of Relativity: Philosophy in Physics 1915–1925*. Oxford: Oxford University Press. A discussion of Cassirer's account of relativity in comparison with other philosophers, like Reichenbach, Schlick, and Eddington.

——(2015). "A Retrospective View of *Determinism and Indeterminism in Modern Physics*." In *The Philosophy of Ernst Cassirer*, edited by J. Tyler Friedman and Sebastian Luft, 65–102. Berlin: de Gruyter. A discussion of the content and reception of DI.

——(2018). "Cassirer and Dirac on the Symbolic Method in Quantum Mechanics: A Confluence of Opposites." *Journal of the History of Analytic Philosophy* Special Issue: Method, Science, and Mathematics: Neo-Kantianism and Early Analytic Philosophy, edited by Lydia Patton and Scott Edgar, 6(3): 213–243. A discussion of DI in general and of the relationship between Cassirer and Dirac, in particular.

Five

Philosophy of culture as the philosophy of symbolic forms

Introduction

Cassirer announces his magnum opus, The Philosophy of Symbolic Forms, with the declaration, "The critique of reason becomes a critique of culture" (PSFv1 9/9/80). With these words, Cassirer at once situates his philosophy of culture in the critical tradition of Kant and indicates his intention to step beyond Kant (and his Marburg predecessors).[1] For though Cassirer defends his philosophy of culture within a critical framework, on his view, it is not 'reason', but rather the more expansive notion of 'symbolic form' that is the cornerstone of culture. Cassirer's system of culture is thus a critical philosophy of symbolic forms. And in developing this system, Cassirer offers an account not only of the cultural world around us, but also of who we are as 'symbolic animals' who create this very world (EM 26/31).

My aim in the next three chapters is to elucidate Cassirer's critical philosophy of culture as a philosophy of symbolic forms. I begin in this chapter by exploring the general philosophical framework that Cassirer develops for his philosophy of symbolic forms. I then turn to an analysis of the individual symbolic forms in Chapter Six (myth, religion, art, language, history, technology, mathematics, natural science) and Seven (right [Recht]).

Before taking up this systematic investigation, I shall make a few remarks concerning the historical development of Cassirer's thought on culture. In much of my examination of Cassirer's philosophy of culture, I take my cue from his early formulations of it in The Philosophy of *Symbolic Forms* and other texts from the 1920s, like "The Concept of Symbolic Form in the Construction of the Human Sciences" (1923) and "The Problem of the Symbol and Its Place in the System of Philosophy" (1927). Even during this period, Cassirer's thoughts on culture evolve. To note one important example, which I return to below, in the first two volumes of *The Philosophy of Symbolic Forms*, Cassirer orients his systematic analysis of culture around three stages of culture: the mimetic, analogical, and symbolic stages. However, in the "Problem of the Symbol," he introduces a new triad, that of expression (*Ausdruck*), presentation (*Darstellung*), and pure significa-tion (*reine Bedeutung*), which serves as the systematic organizing prin-ciple for the third volume.

Cassirer's thought on culture also continues to evolve after writing *The Philosophy of Symbolic Forms*. Two developments, in par-ticular, are of note. First, in *An Essay on Man*, Cassirer chooses to frame his philosophy not as a 'critique of culture' as he does in *The Philosophy of Symbolic Forms*, but rather as a 'philosophical anthro-pology'.[2] The term 'philosophical anthropology' sometimes refers to a tradition in 20th century European philosophy, whose proponents include Max Scheler and Helmuth Plessner; however, in *An Essay on Man*, Cassirer uses the term in a broader sense to refer to a tradition in philosophy that traces back to Plato and Aristotle, which is oriented around the question, 'what is man?' (see EM Ch. 1). Although in pursuing this question Cassirer by no means gives up on his critical approach to culture, he does shift his focus toward an investigation of human nature. And within the frame-work of philosophical anthropology, Cassirer defends the claim that a human being is, most fundamentally, not an "*animal rationale*," as Aristotle would have it, but rather an "*animal symbolicum*" who builds a cultural world (EM 26/31).

The second major development of note is that after the publi-cation of *The Philosophy of Symbolic Forms*, Cassirer develops his phil-osophy of culture in new ethical and political directions. This is not to say that ethical and political considerations were absent in his early work. In *Freedom and Form* (1916) he mounts a defense of a liberal approach to ethics and politics from the perspective of the German Enlightenment. And, as we shall see, he orients

The Philosophy of Symbolic Forms around an account of how culture progresses towards freedom. However, in the 1930s and 1940s he expands the ethical and political dimensions of his philosophy both in the framework of symbolic forms, for example, in *Axel Hägerström* (1939), and in light of the rise of fascism, for example, in *The Myth of the State* (1946).

Yet, in spite of these developments, Cassirer persists in describing his philosophy of culture as a philosophy of symbolic forms. In *An Essay on Man* and *The Myth of the State*, he labels his philosophy of culture in precisely these terms and he cashes out regions of culture as 'symbolic forms' (EM 67–68/75, see also EM 26/31, 143/155, 217/233; MS 34/37). And in *An Essay on Man* he glosses his thesis in philosophical anthropology, viz., that the human being is a symbolic animal, as follows: "Reason is a very inadequate term with which to comprehend the forms of man's cultural life in all their richness and variety. But all these forms are symbolic forms" (EM 26/31).

Thus, even though Cassirer's views on culture evolve in important directions after *The Philosophy of Symbolic Forms*, he regards these as developments of his initial philosophy of symbolic forms. For this reason, I orient my discussion of his philosophy of culture around the philosophy of symbolic forms he defends in the eponymous volumes. However, in doing so, I by no means intend to downplay the significance of the development of his thought. My analysis of his philosophy of culture is grounded in the presupposition that it is, at the same time, an account of the human being as a 'symbolic animal'. And I devote Chapter Seven to an analysis of the ethical and political development of his philosophy of culture.

Preliminaries aside, in this chapter my goal is to offer an overview of Cassirer's framework for culture in general. I begin with a discussion of Cassirer's methodological commitments. As we shall see, although Cassirer continues to deploy the transcendental method of Marburg Neo-Kantianism, albeit in his modified form, he also models his approach to culture on Hegel's 'phenomenology of spirit' and Natorp's bi-directional account of the 'objective' and 'subjective' dimensions of culture.

In the second part of this chapter, I explore the general framework for culture that Cassirer develops on this methodological basis. To this end, I look at Cassirer's account of the unity of culture as something rooted in the free activity of 'spirit' [*Geist*] and as organized around basic 'categories' of spirit. I next turn to Cassirer's account of the 'subjective' dimensions of culture, that is, how it manifests in consciousness and lived-experience. And I survey the 'general' function of 'representation' (*Repräsentation*) that Cassirer takes to orient consciousness as a whole, and the three specific functions which orient it in particular directions, viz., the functions of 'expression' (*Ausdruck*), 'presentation' (*Darstellung*), and 'pure signification' (*reine Bedeutung*). In the last section I take up Cassirer's analysis of the 'objective' dimensions of culture, that is, how it manifests in our cultural world as a whole and in symbolic forms, like myth, language, art, and mathematics. After offering a general account of how Cassirer defines a symbolic form, I address how he organizes his system of symbolic forms around a parallel between them and the functions of consciousness. I then analyze his account of the relation between the symbolic forms. I orient this discussion around two theses: the 'irreducibility thesis', according to which each symbolic form is autonomous, and the 'teleology thesis', according to which there is a teleological progression among the symbolic forms. Once this general framework is in place, I provide an account of the individual symbolic forms in Chapters Six and Seven.

A critique of culture

As was the case with his early philosophy of mathematics and natural science, the transcendental method orients Cassirer's approach to culture. He makes this clear in the following summary remarks:

> [We] attempt to understand the facts . . . But that does not mean that [they] can be deduced in a mere *a priori* manner in thought. We have no other way to find [the facts] than to ask the special sciences, and we have to accept the data with which we are provided by them, the data of the history of language, the history of art, and the history of religion. But what we are searching for are not the historical phenomena themselves. We try to analyze and to understand the fundamental

modes of thinking, of conceiving, representing, imagining, and picturing that are contained in language, myth, art, religion, and even in science. . . . [W]e inquire into the nature of the different functions on which the phenomena, taken as a whole, depend.

(CIPC 80–81/110, see also CP 56–57/148–151)

Adhering to Marburg procedure, Cassirer indicates that he takes as his starting point the 'facts' of culture, which are studied in the 'special sciences'. Though the facts of mathematics and the natural sciences (Naturwissenschaften) continue to be of interest to Cassirer, he also broadens his scope to include the facts pertaining to other regions of culture, like myth, religion, language, and art, which are investigated through the 'human sciences' (Geisteswissenschaften) or 'cultural sciences' (Kulturwissenschaften), including history, art history, anthropology, and linguistics.[3] And just as his writings on mathematics and natural science are marked by a careful engagement with the work of mathematicians, physicists, and chemists, so too are his writings on culture replete with references to research from these disciplines. Indeed, as I emphasized in Chapter One, throughout the 1920s, Cassirer participated in the interdisciplinary community that gathered around the Warburg Library in Hamburg, and the significance of his collaboration with researchers in these other fields is evident throughout The Philosophy of Symbolic Forms and his other writings on culture in the 1920s.

This said, since the transcendental method is not just a method for gathering facts, but rather a method for investigating the conditions that make those facts possible, Cassirer's philosophy of culture is ultimately organized around an effort to elucidate the conditions of culture. Ever in a functional vein, Cassirer glosses these conditions in terms of the different 'functions' that make culture, and the regions within it, possible. However, as I noted in Chapter Two, in his philosophy of culture he comes to think of these functions as functions of 'spirit' [Geist]. For example, he claims that in The Philosophy of Symbolic Forms,

the Copernican revolution, with which Kant began, takes on a new and wider sense. It no longer refers only to the logical function of judgment but extends, with equal justification and right, to every tendency and every principle of spiritual configuration [*geistiger Gestaltung*]. . . . [T]he basic principle of critical thinking, the principle of the "primacy" of the function over the object, assumes in each special domain a new shape [*Gestalt*]. . . . With this, the critique of reason becomes a critique of culture. It seeks to understand and demonstrate how the content [*Inhalt*] of culture . . . insofar as it is grounded in a general principle of form, presupposes an original act of spirit [*Tat des Geistes*]. Herein the basic thesis of idealism finds its true and complete confirmation.

(PSFv1 8–9/9/79–80).

In this passage, Cassirer insists that his philosophy of culture belongs to the tradition of 'critical thinking' and 'critical idealism' because he is guided by the 'primacy of function over the object'. However, he also casts this 'function' not in intellectual terms as he did in *Substance and Function*, but rather in terms of an 'original act of spirit'. Recall that, for Cassirer, this 'spirit' is not to be understood in metaphysical terms as an absolute substance, but rather in terms of an *a priori*, intersubjectively shared structure and activity, which unites human beings, qua human beings, together. On Cassirer's view, it is to the functions of spirit that we must turn in order to understand the conditions of the possibility of culture.

It is for this reason that Cassirer frames his critical idealism and critical epistemology in his philosophy of culture in terms of the functions of spirit. We see him make this move in the passage above, where he claims that the "basic thesis of idealism finds its true and complete conformation" in the recognition that the functions of spirit are the ground of all the objects of culture. Meanwhile, emphasizing the implication the shift towards functions of spirit has for his critical epistemology, Cassirer claims,

When I attempted to apply my findings [from *Substance and Function*] . . . to the problems that concerned the *human sciences* [*Geisteswissenschaften*], it became increasingly clear to me that the general theory of cognition [*Erkenntnistheorie*] . . . would need to be fundamentally broadened. Rather than investigating only the general presuppositions of the scientific *cognizing* [*Erkennens*] of the world, it was equally necessary to differentiate the different basic forms of "*understanding*" [*Verstehen*] of the world and apprehend each one of them as sharply as possible in their distinctive tendency and spiritual [*geistigen*] form.

(PSFv1 lxxix/vii/69, emphasis in original)

Making a similar point in the third volume, Cassirer asserts,

The philosophy of symbolic forms is not concerned exclusively or even primarily with the purely scientific, exact world-comprehending [*Weltbegreifen*] but rather with all the tendencies of *world-understanding* [*Weltverstehens*]. It seeks to apprehend these tendencies of world understanding in their diversity, in their totality, and in the inner differentiation of their manifestations. And it becomes apparent that the "understanding" [*Verstehen*] of the world is no mere taking up, no repetition of a given structure [*Gefüges*] of reality, but instead comprises a free activity of spirit [*freie Aktivität des Geistes*]. There is no true understanding of the world [*Weltverständnis*] that is not in this way based on certain basic tendencies . . . of spiritual *forming* [*geistigen Formung*].

(PSFv3 14–15/14/13, emphasis in original)

In these passages, Cassirer indicates that once he set his sights on culture, it became clear that he would need to expand his critical epistemology beyond an account of mathematical-scientific cognition in order to include the 'understanding' involved in other regions of culture, like myth, religion, language, and art. And it is in this more expansive form that the functions of the 'free activity of spirit' and 'spiritual forming' come to play the key role in his critical epistemology.

Thus, even though in his philosophy of culture, Cassirer continues to defend a version of critical idealism and a critical epistemology, he now casts the project in terms of an account of the functions of spirit that underwrite culture as a whole.

Methodological modifications: the influence of Hegel and Natorp

Although Cassirer's basic outlook on culture is shaped by his long-standing commitment to the transcendental method in general, his approach to culture also betrays the influence of two more specific methodologies for investigating culture, viz., Hegel's conception of a 'phenomenology of spirit' and, what I shall call, Natorp's 'bi-directional' execution of the transcendental method.

Hegel and the phenomenology of spirit

Though Cassirer presents *The Philosophy of Symbolic Forms* in Kantian terms as a 'critique of culture', he also presents it in Hegelian terms as a 'phenomenology of culture'. Indeed, the subtitle of the third volume is "The Phenomenology of Cognition," and Cassirer uses the Introduction to clarify the significance of Hegel's conception of the phenomenology of spirit for *The Philosophy of Symbolic Forms*. There, Cassirer makes clear that by describing his project as a kind of 'phenomenology' he has in mind not the phenomenology of Husserl and Heidegger, but the phenomenology of Hegel, which he describes as follows:

> For Hegel, phenomenology was the basic prerequisite of philosophical cognition because he insisted that philosophical cognition must encompass the totality of spiritual forms [*geistigen Formen*] and because according to him this totality can be made visible only in the transitions from one form to another. The truth [*Wahrheit*] is the "whole"—this whole, however, cannot be given all at once but instead must be unfolded progressively by thought in its own independent movement and rhythm.
>
> (PSFv3 xxxiii–xxxiv/viii/xiv)

Here, Cassirer characterizes Hegel's phenomenology in terms of two, related, sets of methodological demands on a philosophy of culture. On the one hand, a phenomenology of culture requires a systematic accounting of the unity of the cultural world. On the other hand, it demands an account of the development of culture through these cultural forms. And, as Cassirer notes, on the Hegelian view, it is by pursuing this phenomenology through the lens of the 'spirit' that we can offer a systematic and developmental account of culture.

Taking his cue from Hegel, Cassirer presents what he is doing in each volume of The Philosophy of Symbolic Forms as a kind of 'phenomenology'. He presents Volume One as a 'phenomenology of linguistic thinking'; Volume Two as a 'phenomenology of myth', 'mythical thinking', and 'mythical consciousness'; and Volume Three as a 'phenomenology of cognition'.[4] And he continues to describe his project in these terms, for example, in An Essay on Man he frames what he is doing as a "phenomenology of human culture," and in Axel Hägerström he offers a "phenomenology of ethical consciousness" and references a "phenomenology of art" (EM 52/58, emphasis in original; AH 79, 112).[5] In these various contexts, we see Cassirer attempt to execute the systematic and developmental demands of Hegel's phenomenology of spirit.

This said, though Cassirer is drawn to Hegel's phenomenological methodology, he is critical of Hegel's approach to spirit. Hegel organizes his phenomenology of spirit around a tripartite investigation of 'subjective spirit', which encompasses the consciousness of individuals; 'objective spirit', which encompasses the social institutions of right (Recht), ethics, and morality; and 'absolute spirit', which encompasses art, religion, and philosophy. Though there is much debate about how to interpret Hegel's theory of absolute spirit, Cassirer interprets it along metaphysical lines as an "absolute . . . substance" (CIPC 89/117).[6] And given Cassirer's rejection of substance-based metaphysics of any kind, he regards Hegel's appeal to absolute spirit as a misstep.

Instead of turning to Hegel, Cassirer turns to Dilthey in formulating his conception of spirit.[7] Also critical of, what he sees as, Hegel's metaphysical appeal to absolute spirit, Dilthey argues that all of our social and cultural activities should be understood as part of 'objective spirit', which he glosses in terms of the collective, historically unfolding effort of human beings to build a shared cultural world. Persuaded by this line of thinking, Cassirer frames his analysis of the various regions (or 'symbolic forms') of culture as an analysis of 'objective spirit'.[8] He, for example, claims that, "Language, myth, and theoretical cognition are all taken as basic *shapes* [*Gestalten*] of 'objective spirit'" (PSFv3 56/54/49, emphasis in original). And he identifies the "world of human culture" with "the domain of '*objective spirit*'" (PSFv3 316/322/277, emphasis in original; see also PSFv2 xxx/xi/xiv; v3 43/44/39, 65/63/57, 77–78/74/57, 102/99/88). So, even though Cassirer endorses Hegel's methodological approach to culture as a phenomenology of spirit, Cassirer conceives of it as a phenomenology of objective spirit in Dilthey's sense, that is, a phenomenology of the collective, dynamic attempts made by human beings to construct a cultural world.

Natorp and the bi-directional conception of the transcendental method

Cassirer's methodological approach to culture also owes a great deal to Natorp and his 'bi-directional conception' of the transcendental method.[9] According to Natorp, in order to do full justice to the facts of culture, we must refine the transcendental method in order to use it to make sense of both the 'objective' manifestations of culture in art, religion, mathematics, etc., and the 'subjective' manifestation in the consciousness and lived-experience (*Erlebnis*) of individual human beings.[10]

Natorp frames his bi-directional approach to culture in terms of the idea that we can approach culture in a 'plus' or 'minus' direction (Natorp 1912: 71). In the 'plus' direction, the task is to clarify the conditions of the possibility of the objective side of culture. And to do this, Natorp claims that we should rely on a 'constructive' method that elucidates the laws on the basis of which the objects of culture are constructed. Meanwhile, in the 'minus' direction, the goal is to elucidate how culture manifests subjectively in consciousness and lived-experience. However, on Natorp's view, unlike the objects of culture that are directly given to us, consciousness and lived-experiences are never immediately given to reflection. Thus, in order to understand the subjective side of culture, Natorp argues that we need to employ a 'reconstructive' method in which we reconstruct the psychic acts, processes, and operations involved in consciousness and lived-experience. For Natorp, in order to pursue the transcendental method with respect to the objective and subjective dimensions of culture as a whole, we must thus pursue it in the 'plus' direction by means of a more specific constructive method, and in the 'minus' direction by means of a more specific reconstructive method.

Cassirer's commitment to deploying Natorp's bi-directional method is something that he makes explicit in the third volume of *The Philosophy of Symbolic Forms*, specifically in the chapter titled "Subjective and Objective Analysis" (PSFv3 51–66/49–63/45–57). To be sure, in this context, Cassirer criticizes Natorp along the lines I detailed in Chapter Two, viz., for offering an overly rationalist and restrictive account of the subjective and objective dimensions of culture (see Chapter Two). To this end, Cassirer objects that it is not the reason-based notion of 'law', but rather the more encompassing notion of 'form' that we should deploy to make sense of the conditions of culture as a whole (PSFv3 64–65/62–63/56–57). Moreover, Cassirer claims that Natorp neglects important regions of culture, like myth and language (PSFv3 65/63/57).

Yet, in spite of these criticisms, Cassirer embraces Natorp's bi-directional method:

> If we [i.e., in *The Philosophy of Symbolic Forms*] are to obtain a truly concrete view of the "full objectivity" of spirit on the one hand and of its "full subjectivity" on the other, we must seek to carry out the methodological correlation, which Natorp sets forth, in principle, in every domain of spiritual creating [*Schaffen*].
>
> (PSFv3 65/63/57)

Emphasizing the importance of pursuing the subjective direction in particular, Cassirer continues,

> [the philosophy of symbolic forms] aspires to find its way back to the primary subjective "sources" [*Quellen*], the original modes of conduct [*Verhaltungsweisen*] and the original modes of the configuration [*Gestaltungsweisen*] of consciousness. It is from this perspective that we now approach our question: the question of the structure of perceptive, intuitive, and cognitive consciousness. We shall

attempt to elucidate it without surrendering to the method . . . of natural-scientific, causal-explicative psychology. . . . We instead start from the problems of "objective spirit," from the gestalts [*Gestalten*] in which it consists . . .; however, we will not stay with them as a mere factum [*Faktum*] but instead attempt through a reconstructive analysis to find our way back to their elementary presuppositions, to the "conditions of their possibility."

(PSFv3 65/63/57)

With these words, Cassirer clarifies a key component of his agenda in the third volume of *The Philosophy of Symbolic Forms*, viz., pursuing the transcendental method in the 'minus' direction in order to clarify the 'structure of perspective, intuitive, and cognitive consciousness', which underwrites the subjective dimensions of culture.

Appreciating the influence of Natorp is crucial for understanding the third volume of *The Philosophy of Symbolic Forms*, which can, in some ways, seem anomalous. Whereas in the first two volumes of *The Philosophy of Symbolic Forms*, Cassirer concentrates on the objective developments of language, myth, and religion, in the third volume he blends together objective considerations about mathematics and natural science with subjective considerations about structures of consciousness. However, once we read the third volume in light of his endorsement of Natorp's bi-directional method, Cassirer's strategy in the third volume makes sense: he aims to provide a complete account of culture, which accommodates both its objective and subjective dimensions.

Stepping back, although in his philosophy of culture Cassirer continues to pursue the transcendental method, his method is also shaped by Hegelian and Natorpian commitments. In the Hegelian vein, Cassirer endeavors to offer a phenomenology of objective spirit, which does justice to the systematic and developmental features of culture. And in the Natorpian vein, he attempts to give a bi-directional account of how culture manifests 'objectively' in the world around us and 'subjectively' in consciousness and lived experience. Given this complex methodological framework, it should come as no surprise that Cassirer's philosophy of culture is multi-layered. In order to do justice to the transcendental method, Cassirer's philosophy of culture involves careful interdisciplinary engagement with research in the 'natural' and 'human' sciences. He also seeks to offer an account of the unity and diversity of culture, as it manifests subjectively in consciousness and lived experience and objectively in the world around us. In order to elucidate this latter feature of his philosophy of culture, in the next section I discuss his account of the underlying unity of culture as a whole, insofar as it is grounded in spirit. I then turn to his systematic account of the subjective and objective manifestations of culture.

The unity of spirit

Although in his philosophy of culture Cassirer directs much of his efforts toward giving an account of the different ways in which culture manifests, he also defends a theory of the underlying unity of culture. As I anticipated above, on Cassirer's view, the key to understanding the unity of culture is understanding the activity and functions of spirit on which culture is based. For whether we consider the subjective or objective dimensions of culture, Cassirer claims that they have an *a priori* ground in the activity and functions of spirit.

Cassirer offers a basic characterization of the activity of spirit in a passage I cite above:

> it becomes apparent that the "understanding" [*Verstehen*] of the world is no mere taking up, no repetition of a given structure [*Gefüges*] of reality, but instead comprises a free activity of spirit [*freie Aktivität des Geistes*]. There is no true understanding of the world [*Weltverständnis*] that is not in this way based on certain basic tendencies . . . of spiritual forming [*geistigen Formung*].
> (PSFv3 14–15/14/13, emphasis in original)

Here, Cassirer implicitly rejects a substance-based account of the activity of spirit. On the substance-based view, the activity of spirit involves us being receptive to a reality that is brutely given to us. By contrast, Cassirer treats the activity of spirit as a 'free' activity in which we 'form' reality in the first place. In keeping with his critical idealism, Cassirer thus treats the free activity of spirit as something that reality 'conforms to'. However, on his view, the 'reality' that conforms to our spiritual activity is not just the reality of the world around us, but also the subjective reality of consciousness. For Cassirer, then, the activity of spirit is the condition of possibility of the subjective and objective dimensions of culture. Our consciousness and lived-experience, as much as our cultural world are grounded in the *a priori* activity of 'spiritual forming'.

Cassirer, in turn, claims that there are fundamental 'functions' of spirit by means of which this 'forming' takes place:

> To apprehend the laws of this spiritual forming [Formung], we had above all to distinguish their different dimensions sharply from each other. Certain concepts—such as those of number, time, and space—constitute, as it were, originary-forms [Urformen] of synthesis that are indispensable wherever a "multiplicity" is to be taken together into a "unity," wherever a manifold is to be divided and organized according to certain shapes [Gestalten].
>
> (PSFv3 15/14–15/13)

Here, Cassirer further clarifies the activity of spiritual forming as one in which we bring 'unity', 'division', and 'organization' to a 'manifold', that is, to a multiplicity or diversity, either outside of us or in consciousness. And he indicates that there are certain fundamental 'concepts', like 'number, 'time', and 'space', that are 'indispensable' for this 'forming'. As we might cast this point, using language from *Substance and Function*, Cassirer conceives of these 'concepts' as 'universal invariants' of the activity of spirit: they are the concepts that remain constant across all our spiritual activities because they are *a priori* conditions that make spiritual 'forming' possible in the first place. I shall refer to these as the 'categories of spirit'.

Cassirer presents a more elaborate list of the categories of spirit in the first volume of *The Philosophy of Symbolic Forms*,

> If we attempt to set out before us an initial general overview of the totality [Gesamtheit] of relations by which the unity of

consciousness is designated and as such constituted, then we are first led to a series of certain basic relations [Grundrelationen] that confront one another as distinctive and autonomous "modes" of connection [Verknüpfung]. The element of "juxta-position" [Nebeneinander] as it appears in the form of *space*; the element of "succession" [Nacheinander] as in the form of *time*; and the connection of determinations of being in such a way that one is grasped as a "*thing*," the other as a "*property*," or of successive events [Ereignissen] in such a way that the one appears as the *cause* of the other are all examples of such original types of relation.

(PSFv1 25/25–26/95, emphasis in original)

So, in addition to 'number', 'time', and 'space, Cassirer identifies 'thing', 'property', and 'cause' as 'basic relations', which condition and remain invariant across all activity of spiritual formation.

Many of these categories should sound familiar from my discussion of Cassirer's theory of the categories in Chapter Four; however, in The Philosophy of Symbolic Forms, Cassirer modifies this earlier theory of categories in several ways. As mentioned above, instead of framing these categories in intellectual terms, he frames them in terms of spirit. Furthermore, he no longer treats these categories as the categories of mathematical-scientific cognition alone; he treats them as categories that underwrite culture as a whole. Finally, in order to do justice to the pervasive role the categories play across cognition and understanding, Cassirer articulates some of the categories in more generic terms. He replaces the category of 'magnitude' [Größe] with the category of 'thing', and the category of the 'functional dependency of magnitudes' with the category of 'causality'. Nevertheless, Cassirer remains committed to the idea these categories are 'universal invariants' of spiritual activity: they are *a priori* conditions of the possibility of spiritual 'forming'.

Moreover, as in Substance and Function, in The Philosophy of Symbolic Forms Cassirer argues that we need to distinguish between a stable and dynamic moment of the categories of spirit, which he refers to in terms of their 'quality' and 'modality' (PSFv1 26/27/95). This quality–modality distinction maps onto the distinction between logical form and content that I discussed in Chapter Four. Thus, the 'quality' of a category amounts to the basic logical relation that each category involves. For example, the quality of space is the relation of juxtaposition (*Nebeneinander*) and time is the relation of succession (*Nacheinander*) (see PSFv1 26–27/27/95–96). And, on Cassirer's view, the quality of the category is something that remains invariant across all our cultural activities.

By contrast, Cassirer claims that the 'modality' of the categories is something that is dynamic:

> If we designate the different kinds of relation [*Relation*]—such as the relation of space, time, causality, etc.—as R1, R2, R3 . . ., then to each one there belongs a special "index of modality," $\mu1$, $\mu2$, $\mu3$, indicating the function- and signification-context [*Funktions- und Bedeutungszusammenhangs*] in which it is to be taken. (PSFv1 28/29/97, translation modified)[11]

The 'modality' of the categories thus serves as the counterpart of the content of the categories, which changes from scientific theory to scientific theory. However, in the cultural framework, Cassirer makes the more generic claim that the 'modality' of the categories is 'indexed' to a certain 'context'. The 'context' he has in mind is a semantic context in which we 'form' things by unifying, ordering, and organizing them in meaningful ways. And he maintains that we find such contexts operative in both the subjective and objective domains of culture.

As I shall discuss at more length below, subjectively, he maintains that there are certain contexts in consciousness, which orient us towards lived-experiences that are more 'subjective', 'objective', or 'ideal' in character. He describes these subjective contexts as 'functions of consciousness'. And objectively, he argues each region of culture, or symbolic form, serves as a context: "each of these contexts of signification [*Bedeutungszusammenhangs*], language as well as scientific cognition, art as well as myth, possesses its own constitutive principle that imprints its signet, as it were, on all the particular configurations [*Gestaltungen*] within it" (PSFv1 28/29/97). His basic idea is that although all of these subjective and objective contexts involve activities that are made possible by the categories, the 'modality' of a category is something that shifts depending on the context in which it operates. For example, in consciousness, the temporal relation of 'succession' takes on a different modality if we are experiencing everything through a 'subjective' or an 'objective' lens, and the causal relation of 'origination' takes on a different modality in the context of myth than in the context of natural science. Thus, even though Cassirer regards the categories of spirit as relations that remain invariant across all spiritual activity, he conceives of these invariants as having a stable quality and dynamic modality.

Ultimately, with his account of the free activity and categories of spirit, Cassirer offers an account of the underlying unity of culture as a whole. On his view, the free activity and categories of spirit serve as the *a priori* ground of the subjective and objective dimensions of culture. And though, as we shall now see, he devotes much of his energy to elucidating the diverse ways in which this activity takes shape in each context, he is guided throughout by the idea that all our lived-experience and world-building is made possible by this unity of spirit.

The subjective dimensions of culture

With this overarching picture of unity in place, we can now turn to Cassirer's attempt to offer a bi-directional account of culture, which accounts for the subjective and objective dimensions of culture. In this section, my focus is on Cassirer's account of the subjective dimensions of culture, that is, his characterization of the consciousness and lived-experience of human beings who are embedded in a cultural world. And in the following section, I explore his analysis of the objective dimensions, considering the ways in which he attempts to fill out his 'phenomenology of spirit' therein.

In elucidating the subjective dimensions of culture, Cassirer characterizes consciousness and lived-experience in terms of 'functions'; however, these functions are not the same as the categories I have just discussed. The functions of consciousness are what provide the context in which the categories receive a distinctive 'modality'. In order to explore his account of the functions of consciousness, I turn first to his account of the 'general' function that orients consciousness as such and three more 'specific' functions of consciousness that orient it in particular directions.

The general function of consciousness

According to Cassirer, in the context of consciousness, our spiritual activity manifests as a way of bringing unity to a 'manifold' in consciousness. Cassirer describes this basic conscious activity as the activity of 'representation'. He makes this clear in a section of the first volume of The Philosophy of Symbolic Forms, titled "The Problem of 'Representation' [Repräsentation] and the Construction [Aufbau] of Consciousness" (PSFv1 24–38/25–39/93–105). There, he argues that consciousness in general is grounded in a function that he calls the "function of representation [Repräsentation]" (PSFv1 31/32/99).[12]

On Cassirer's view, 'representation' involves "the presentation [Darstellung] of one content in and through another" (PSFv1 38/39/104). Or, as he describes representation later, "every separate content of consciousness is situated in a network of manifold relations, by virtue of which . . . it also includes in itself a pointing to [Hinweis] other and still-other contents" (PSFv1 39/40/106, emphasis in original). By a 'content of consciousness', Cassirer has in mind something we are conscious of. Thus, on his view, representation in consciousness amounts to taking one thing we are conscious of as 'pointing to' something else we are conscious of, for example, when I am conscious of the A-flat I hear as 'pointing to' the melody I am listening to.

According to Cassirer, this general function of representation "should be recognized as an essential prerequisite for the construction [Aufbau] of consciousness itself and as a condition of its inherent formal unity" (PSFv1 38/39/104). He thus conceives of the basic activity of consciousness as one in which we unify a manifold in consciousness vis-à-vis representational relations. And as an activity that has its source in spirit, Cassirer thinks that representation is made possible by the categories of spirit. To make this point using language I cited above, on his view, the categories are the "originary-forms [Urformen] of synthesis" through which we representationally relate one thing we are conscious of to another (PSFv3 15/14–15/13). In the context of consciousness in general, the categories thus provide the basic framework for representation.

The specific functions of consciousness

In addition to this general discussion of consciousness, Cassirer details three more specific contexts in consciousness, which involve distinctive modes of representation. He indexes each of these contexts and modes of representation to a specific function of consciousness: the functions of expression (*Ausdruck*), presentation (*Darstellung*), and pure signification (*reine Bedeutung*) (see Table 5.1 below).

According to Cassirer, the function of expression provides a context that is more subjective in character. In this context, we represent things as having affective or 'physiognomic' characteristics (PSFv3 79/76/68). For example, when I am conscious of someone's face as joyful or a daffodil as triumphant, this occurs in the context of expression.

Cassirer labels the mode of representation that accompanies the function of expression 'perception' (*Wahrnehmung*), or, more precisely, 'expressive perception' (*Ausdruckswahrnehmung*) (see PSFv3 85/82/74, 91–92/88–90/78–80). Thus, in the expressive context, we represent one content of consciousness as 'pointing to' another by perceiving the one content as expressed in the other. For Cassirer, it is crucial to understand this as a kind of perception. Under the guidance of the function of expression, we represent one content in and through another by perceiving it as immediately expressing that other content. So, in order to expressively represent joy on your face, I do not need to infer joy from your face; I can immediately see it as expressed in your face.[13]

By contrast, Cassirer claims that the function of presentation provides a context that is more objective in nature. In this context, we are conscious of objects, properties, and events, which are independent from our affects, desires, or other subjective states. For example, when I see a daffodil through the lens of the function of presentation, instead of anthropomorphizing it, I am conscious of it as an object that exists independent from me.

Cassirer describes the mode of representation that belongs to the function of presentation as 'intuition' (*Anchauung*) or 'thing perception' (*Dingwahrnehmung*) (see, e.g., PSFv3 91/88/79, 135/130/117, 139/132/119). Under the guidance of the function of presentation, we grasp one content of consciousness as 'pointing to' another by intuiting or perceiving one content in the other. And, as was the case with expressive perception, by describing this as a kind of 'intuition' or 'perception', Cassirer wants to emphasize that the representation mediated by the function of presentation has the character of immediacy. In my intuition of the daffodil as an object, I do not judge it to be independent from me; I see it as independent from me. Nevertheless, Cassirer regards this kind of perception as distinct from expressive perception because it involves consciousness of an objective order that is independent from subjects.

According to Cassirer, the third function of pure signification provides a context that is ideal in its orientation. As he puts it, this function orients consciousness toward the "the realm of pure signification [*reinen Bedeutung*]," which includes 'pure significations', like mathematical functions, logical laws, or moral principles (PSFv3 334/326/284). Thus, when I am conscious of a wavy line as a sine function, on Cassirer's view, I am conscious of the line as representing a pure signification (PSFv3 334/326/284).

Cassirer characterizes the mode of representation that accompanies the function of pure signification as 'cognition' (*Erkenntnis*). And he contrasts cognition with expressive perception and intuition as follows:

> There now develops a kind of detachment, of "abstraction" that was unknown to perception and intuition. Cognition releases the pure relations from their involvement with the concrete and individually determined "reality" of things, in order to re-present [*vergegenwärtigen*] them purely as such in the universality of their "form."
>
> (PSFv3 334–335/326/284)

According to Cassirer, whereas in expressive perception and intuition our consciousness is immersed in what is sensuously present, in cognition we 'detach ourselves' from what is sensuously present by representing something we are conscious of as 'pointing to' a pure signification. Thus, when I cognize the line as a sine function, I am conscious of the line as representing a 'pure' signification that is ideal, rather than sensuous in character.

Table 5.1 summarizes Cassirer's framework for the specific functions of consciousness.

According to Cassirer, although the representational activity involved in each function of consciousness is made possible by the categories, the categories take on a different 'modality' in each context. For example, our sense of what 'succession' or 'juxtaposition' amounts to in the subjective context of expressive perception is different from the objective context of intuition and the ideal context of cognition. So even though all of our conscious activities involve drawing representational connections within the framework of the categories, the categories are specified in a different way in each of these contexts.

With this analysis of the general and specific functions of consciousness, Cassirer takes himself to have provided an overarching picture of the subjective dimensions of culture, a picture that reflects the consciousness and lived-experience of the human beings who inhabit a cultural world. Though this picture of consciousness sheds light on part of Cassirer's understanding of what it is to be a human being, it does not exhaust his view. For Cassirer, if we are to fully

Table 5.1 The functions of consciousness

Specific function of consciousness	Specific mode of representation
function of expression (*Ausdrucksfunktion*)	perception (*Wahrnehmung*), expressive perception (*Ausdruckswahrnehmung*)
function of presentation (*Darstellungsfunktion*)	intuition (*Anschauung*), thing perception (*Dingwahrnehmung*)
function of pure signification (*reine Bedeutungsfunktion*)	cognition (*Erkenntnis*)

understand the human being, we must also take into account her status as a 'symbolic animal' who constructs a cultural world. It is to these constructive efforts and the objective dimensions of culture which we shall now turn.

The objective dimensions of culture

At the heart of Cassirer's account of the objective dimensions of culture is the notion of a 'symbolic form'. Symbolic forms are, as he puts it, the "paths [*Wege*] by which spirit proceeds in its object-ivization [*Objektivierung*]" (PSFv1 7/7/78). Taken together, Cassirer claims that the symbolic forms make up our shared cultural world as a whole, a world he refers to as our 'symbolic universe', 'common world', or 'world of humanity' (EM 25/30, 202/218, 221/237; CIPC 73/103–104, 90/118; CP 55/149). And it is through his analysis of the system and development of the symbolic forms that Cassirer attempts to elucidate the objective manifestations of our spiritual activity and complete his phenomenology of objective spirit.

While I shall take up Cassirer's account of the individual symbolic forms in the next two chapters, in what remains of this chapter I consider his system of symbolic forms in general. To this end, I explore what Cassirer means by a 'symbolic form', how he connects the symbolic forms to his discussion of the functions of consciousness, and his analysis of how the symbolic forms relate to one another.

'Symbolic form'

So, what exactly is a symbolic form according to Cassirer? One way to answer this question is by referring to Cassirer's examples: a symbolic form is a region of culture, like myth, religion, language, art, mathematics, natural science, or right (*Recht*). Though this is true to an extent, we must be careful in how we understand a region of culture as a 'symbolic form'. Frequently, when we think of a region of culture, we think of the set of cultural objects and practices that fall in that domain. For example, when we think of myth, we think of particular myths, totems, taboos, and rituals. Or when we think of art, we think of particular paintings and poems, as well as practices, like ballet or blues. Cassirer labels these cultural objects and practices 'configurations' (*Gestaltungen*). And though he sees a connection between configurations and symbolic forms, he does not identify the two. He, instead, treats a symbolic form as the source of these configurations.

To this end, Cassirer argues that the form involved in a symbolic form should be understood in active terms. As he helpfully makes this point, a symbolic form is a *forma formans*, an actively forming form, which produces a cultural configuration as a *forma formata*, a formed form.[14] Or as he sometimes puts it, drawing on Humboldt, a symbolic form is the *energeia* that produces the *ergon*, the works, of culture.[15] And, indeed, this makes sense given his conception of a symbolic form as an objective manifestation of the free activity of spirit. Thus, for Cassirer, the objects and practices of culture are to be understood as the products of the '*forma formans*' or '*energeia*' that is a symbolic form.

Cassirer, in turn, describes a symbolic form as 'symbolic' because he thinks that it provides a semantic context in which cultural objects and practices take on meaning. In elucidating this aspect of a symbolic form, Cassirer claims that each symbolic form provides a broad 'context of signification' in which configurations take on meaning: "each of these contexts of signification [*Bedeutungszusammenhangs*], language as well as scientific cognition, art as well as myth, possesses its own constitutive principle that imprints its signet, as it were, on all the particular configurations [*Gestaltungen*] within it" (PSFv1 28/29/97). For example, different words, sentences, or grammatical practices take on meaning within the broad 'context of signification' provided by the symbolic form of language. Notice that Cassirer treats this 'context of signification' as one that encompasses all the configurations within a particular region of culture. On his view, then, regardless of when or where a cultural configuration is produced, it will be semantically connected to every other configuration in that region by means of the context of signification provided by a symbolic form. For example, whether we think of classical Sanskrit, middle French, or modern Japanese, according to Cassirer, these languages are semantically bound together in the broader context of language as a symbolic form.

Cassirer, moreover, describes the meaning that cultural configurations take on as 'symbolic' in character. However, Cassirer has a particular conception of what symbolic meaning amounts to. Symbols are often taken to have an indirect relation, such as, an analogical or metaphorical relation, to what they symbolize. On Cassirer's view, by contrast, there is a more direct relation between a symbol and what it symbolizes that he attempts to capture with the notion of 'symbolic pregnance' (Prägnanz).[16]

'Prägnanz' is a notoriously difficult German word to translate, as there is no precise English equivalent for it. The term has its roots, on the one hand, in the Latin term 'praegnans', which means both 'to expect a child' and 'to be full'.[17] On the other hand, Prägnanz is related to the verb 'prägen', which means 'to shape' or 'to mint'. It is this latter sense of Prägnanz that the Gestalt psychologists draw on in their account of the 'laws of Prägnanz' that govern how we organize the perceptual world. And Cassirer appears to have both connotations in mind in his discussion of symbolic Prägnanz. Thus, on his view, to say that something is symbolically pregnant is to say that it is full of what it symbolizes, imprinted with and shaped by that meaning.

For Cassirer, then, the cultural configurations we produce through the symbolic forms do not just analogically or metaphorically refer to some meaning; they are 'symbolically pregnant' with that meaning. In endorsing this view, Cassirer is guided by a conception of 'matter' and 'form' that I discussed in Chapter Two. Recall that, on Cassirer's view, it is a mistake to think that there is any matter that we are brutely given, which has not been shaped by some form. Applying this to the configurations of culture, Cassirer thinks that our cultural objects and practices should not be understood in terms of some brutely given matter that we attach some symbolic meaning to. We should understand these configurations as 'formed' from the outset: their meaning is inseparable from them. For Cassirer, then, through the symbolic forms we produce objects and practices that are meaningful to their core, and it is these objects and practices that make up our cultural world.

In sum, on Cassirer's view, a symbolic form is something active, a forma formans, by means of which we produce symbolically meaningful objects or practices of culture. We do so in the broad context of signification provided by each symbolic form and each configuration we produce is 'symbolically pregnant' with its meaning.

The system of symbolic forms

With the basic concept of a 'symbolic form' in place, we can now turn to Cassirer's account of how the symbolic forms are systematically woven together in the cultural world as a whole. In order to lay out the contours of this system, I begin with Cassirer's organization of the symbolic forms around a parallel with the functions of consciousness. I then address his analysis of the relation the symbolic forms have to one another.

In order to clarify the 'context of signification' that belongs to each symbolic form, Cassirer draws on his analysis of the functions of consciousness. To this end, he argues that in the symbolic forms we find an objective manifestation of the patterns and orientations that subjectively manifest in the functions of consciousness. He develops this parallel as shown in Table 5.2.

Although Cassirer does not dwell on the parallel between the general function of consciousness and the cultural world at length, the two domains manifest a similar tendency towards constructing a whole. In the subjective context, this whole involves the 'construction' (Aufbau) of consciousness, and this happens by means of each individual continuously connecting the contents of her consciousness together through representational relations (PSFv1 24/25/93). In the objective context, this whole involves the 'construction' of a shared cultural world. And, on Cassirer's view, this construction happens through the collective, historically unfolding efforts of human beings, qua 'symbolic animals', who build a symbolically meaningful world around themselves.

Cassirer devotes much more attention to teasing out the subjective-objective parallels between specific functions of consciousness and individual symbolic forms. As I shall discuss at length in Chapters Six and Seven, he claims that myth and religion are animated by the subjective tendencies operative in the function of expression; art displays these subjective tendencies as well as the objective

Table 5.2 The subjective and objective dimensions of culture

Subjective		Objective
General function of consciousness		*General objective manifestation of spirit*
function of representation		the cultural world as a whole ('symbolic universe', 'common world', 'world of humanity')
Specific function of consciousness	*Specific mode of representation*	*Specific symbolic form*
function of expression (*Ausdrucksfunktion*)	perception, expressive perception	myth, religion
		art
function of presentation (*Darstellungsfunktion*)	intuition, thing perception	language, history, technology
function of pure signification (*reine Bedeutungsfunktion*)	cognition	mathematics, (mathematical) natural science, right (*Recht*)

tendencies manifest in the function of presentation; language, history, and technology recapitulate the objective tendencies of presentation; and mathematics, natural science, and right (*Recht*) manifest the ideal orientation at work in the function of pure signification. For now, though, it will suffice to note that part of Cassirer's system of symbolic forms involves appreciating that the symbolic forms objectively manifest the subjective, objective, and ideal orientations that are subjectively manifest in the domain of consciousness and lived-experience.

In addition to articulating these subjective-objective correlations, with his system of symbolic forms Cassirer seeks to clarify the relation the symbolic forms have to one another. His analysis of these relations turns on two theses, which I shall refer to as the 'irreducibility thesis' and 'teleology thesis'.

According to Cassirer's irreducibility thesis, no symbolic form can be 'reduced to' or 'derived from' any other symbolic form:

> each of [the symbolic forms] creates its own symbolic configurations [*Gestaltungen*], which . . . are . . . equal as to their spiritual origin [*geistigen Ursprung*]. None of these configurations can simply be reduced to, or derived from, the others; rather, each of them designates a determinate mode of spiritual apprehension, in and through which it constitutes its own aspect of the "actual" [*Wirklichen*].
>
> (PSFv1 7/7/78)

On his view, the irreducibility thesis is not to be understood as an empirical claim about the genetic or historical origin of the symbolic forms. Indeed, he maintains that if we consider the historical origin of the symbolic forms, there is a sense in which they all spring from myth:

> None of [the symbolic forms] immediately emerges as a separate, independent, and recognizable configuration, but each gradually detaches itself from the common earth of myth. All the contents of spirit, however much we are able to systematically assign them to their own domain and base them on their own autonomous "principles," are factually first given to us only in this interpenetration. Theoretical, practical, and aesthetic consciousness—the world of language and cognition, art, right [*Recht*], and ethics, the basic forms of the community and the state—all of these are originally bound to mythical-religious consciousness.
>
> (LM 168/266, see also PSFv2 xxx–xxxi/x–xii/xiv–xv)

So too in the case of language, Cassirer acknowledges the ways in which the historical development of the other symbolic forms is bound up with language in some way. Just think, for example, of the pervasiveness of language in mythical rites, religious texts, poetry, prose, law, or scientific theories.

Rather than being an empirical thesis about the genetic or historical origin of the symbolic forms, Cassirer's irreducibility thesis is a 'critical' thesis, according to which the 'function' of each form cannot be reduced to or derived from any other symbolic form (PSFv2 15/15/13). For Cassirer, the function of each form is tied to a distinctive 'context of signification' and mode of 'objectivization' (*Objektivierung*) by means of which we build up our cultural world. And his point with the irreducibility thesis is that the function of each symbolic form involves a function that cannot be reduced to any other function. Thus, on Cassirer's view, even when the symbolic forms are empirically intermingled with one another, they each possess an autonomous function that makes a distinctive contribution to our cultural world.

In defending the irreducibility thesis, Cassirer juxtaposes his position with Hegel's position on the relationship between the regions of culture. On Cassirer's interpretation, Hegel thinks that, "All spiritual being and events [*geistige Sein und Geschehen*] . . . [are] in the end, reduced to and based on a single dimension," viz., the dimension of 'logic' (PSFv1 13/13/84). In Hegel's system, Cassirer claims that, "of all the spiritual forms, only that of the form of the logical, the form of the concept and of cognition, appears to have been ascribed a real, true *autonomy*" (PSFv1 13/13/83, emphasis in original). As Cassirer reads Hegel, then, the other forms of culture, like art, religion, right (*Recht*), are reducible to the form of logic because they are ways in which logic unfolds.

On Cassirer's view, by contrast, the symbolic forms cannot be derived from or reduced to one another because they are autonomous from one another. Thus, instead of one form dominating all the others, Cassirer asserts that the relations between the forms should be understood in terms of 'harmony in contrariety':

> Language, art, religion, science . . . tend in different directions and obey different principles. But this multiplicity and disparateness does not denote discord or disharmony. All these functions complete and complement one another. Each one opens a new horizon and shows us a new aspect of humanity. The dissonant is in harmony with itself; the contraries are not

mutually exclusive, but interdependent: "harmony in contrariety, as in the case of the boy and the lyre."

(EM 228/244)[18]

Here, drawing on Heraclitus's metaphor of a bow and lyre, Cassirer offers a kind of pluralistic account of the relation between the symbolic forms, as co-existing in their difference. And one of the main tasks Cassirer pursues in his phenomenology of objective spirit is elucidating the irreducible features of each symbolic form.

However, in addition to the irreducibility thesis, Cassirer defends the teleology thesis, according to which there is a teleological progression among the symbolic forms. In this vein, Cassirer tends to use another metaphor, viz., that of Hegel's metaphor of a 'ladder' of culture (see, e.g., PSFv2 xxxi–ii/xii–iii/xv–xvi, PSFv3 xxxiv/ix/xv). And in the third volume of The Philosophy of Symbolic Forms, Cassirer defends a view according to which myth and religion belong on the lowest rungs of this ladder, language occupies the middle rungs, and mathematics and natural science occupy the highest rungs.

There is a debate about how to understand these teleological remarks, especially in light of Cassirer's irreducibility thesis. Some commentators take Cassirer to be committed to a hierarchy of symbolic forms.[19] Others argue that his position is ultimately a 'pluralistic' or 'complementarist' one in which all the symbolic forms stand in a 'centrifugal' or 'vertical' relationship to one another.[20] Still others argue that there is a fundamental tension in Cassirer's position between these hierarchical and egalitarian commitments.[21]

In order to navigate this debate, it is important to recognize different senses in which the symbolic forms may, or may not, be 'equal' with one another. In his defense of the irreducibility thesis, Cassirer treats the symbolic forms as 'equal' in the sense that each makes a distinctive contribution to the cultural world. However, in pursuing the teleology thesis, Cassirer has another kind of equality in mind: equality with respect to the end or telos of culture.

Like Kant and Hegel before him, Cassirer conceives of the telos of culture in terms of the "consciousness of freedom," and he, therefore, regards progress in culture as the "progress of consciousness of freedom" (CIPC 90/118). By 'consciousness of freedom', Cassirer has in mind the consciousness of the 'free activity of spirit' that underwrites all of our individual and collective activities. By Cassirer's lights, this activity is 'free' in the sense that it is not determined by anything outside of spirit, but rather involves self-determination through spirit. As such, Cassirer conceives of this freedom as involving spontaneity and autonomy.

In his discussion of how we become conscious of freedom, Cassirer discusses a kind of practical and theoretical recognition of freedom. In a practical vein, he claims that we can be conscious of our freedom by recognizing our status as spontaneous and autonomous agents. Meanwhile, in what I shall describe as a 'theoretical' vein, Cassirer examines ways in which we can become conscious of the free activity of spirit as something that the world conforms to. On his view, this theoretical recognition of freedom involves us realizing that the basic structures of the world, viz., space, time, numbers, things, properties, and causes, are not substances, but rather categories that have an *a priori* or ideal source in the spontaneity and autonomy of spirit. For Cassirer, then, progress towards 'consciousness of freedom' involves progress toward a practical and theoretical recognition of freedom. And when we look at the symbolic forms from this teleological perspective, Cassirer denies that all the symbolic forms stand on equal ground and insists, instead, that some symbolic forms encourage in us a higher degree of consciousness of freedom than do others.

In pursuing his analysis of the teleological progress of the symbolic forms, Cassirer explores the ways in which this progress unfolds both internal to and across the symbolic forms. With respect to his account of internal progress, Cassirer maintains that each of the symbolic forms progresses towards an increased consciousness of freedom through three phases: a mimetic, analogical, and symbolic phase (see PSFv1 133–143/132–146/186–197; PSFv2 288–290/277–280/237–239, 307–313/298–306/254–261; PSFv3 524–525/522–523/450–451; CSF 79–88/82–85). In the mimetic phase, he claims that a symbolic form encourages us to conceive of ourselves as passive with respect to a world that is wholly independent from us. Then in the analogical phase, the symbolic form promotes an understanding of who we are as agents and an understanding of, at least part of the world, viz., cultural configurations, as analogues or representations of a mind-independent reality. Finally, in the symbolic phase, the symbolic form fosters a recognition of who we are not just as agents, but as spontaneous and autonomous agents, and a recognition of the world as a whole as something that conforms to that spontaneity and autonomy. Through each phase, Cassirer maintains that the symbolic form promises us a higher degree of the practical and theoretical consciousness of freedom.

Though Cassirer thinks that each of the symbolic forms is capable of this sort of internal development, he claims that when we look at progress across the symbolic forms it becomes clear that there are limits to how far some of them advance our consciousness of freedom. On his view, the symbolic forms that parallel the functions of expression and presentation remain tethered to the sensuous world and, as a result, they can only promote a consciousness of freedom that is bound to the sensuous world. Thus, even in the symbolic phases of these forms, Cassirer argues that we cannot achieve insight into our freedom as something that has an ideal source in spirit. By contrast, he argues that given the ideal orientation of the symbolic forms that parallel the function of pure signification, they are able to promote a recognition of our freedom as something that has an ideal spiritual origin. It is thus in the symbolic phases of these forms that Cassirer thinks we should look for the most teleologically advanced stage of our practical and theoretical consciousness of freedom. And it is this teleological development across culture, rather than teleological development internal to a symbolic form that I shall focus on in the following chapters.

This said, although Cassirer devotes much of his attention to explicating the progress up the 'ladder' of culture, he denies that this progress is either necessary or guaranteed. This is another topic on which Cassirer takes himself to part ways with Hegel. For, according to Cassirer's interpretation, Hegel conceives of cultural progress as something that is necessary, having been dictated by the "universal plan" of absolute spirit (CIPC 89/117). On Cassirer's view, by contrast, the progress of culture is contingent: although there have been advancements in culture, it could have been otherwise. What is more, in The Myth of the State Cassirer argues that the rise of fascism in the 1930s and 1940s made culture, in fact, be otherwise (see

Chapter Seven). Cassirer thus balances his account of the teleological progress of culture with an analysis of the ways that culture can and does regress.

This recognition of the fragility of cultural progress speaks to the ongoing importance of the irreducibility thesis on Cassirer's view. For even in more teleologically advanced stages of culture, Cassirer maintains that each symbolic form remains an independent and autonomous force within it. This opens up space not only for continued conflict among the symbolic forms, but also for regression if certain forms should gain ascendency over others. For Cassirer, then, the progress of culture is something that, in the end, involves a struggle, a struggle that can, but need not develop toward an increase in consciousness of freedom.

Thus, although in defending the irreducibility thesis, Cassirer highlights the ways in which the symbolic forms are equal, qua forms that make a distinctive contribution to our cultural world, in defending the teleology thesis, he argues that they are not equal with respect to the telos of culture, as some advance the consciousness of freedom more than others. And in the following two chapters, my analysis of the individual symbolic forms will be guided by an effort to balance Cassirer's account of the irreducible features of each symbolic form with his discussion of their teleological progression towards freedom.

Summary

My aim in this chapter was to lay out the general framework of Cassirer's philosophy of culture. I began by exploring his methodology. In addition to discussing Cassirer's pursuit of the transcendental method, I examined his Hegelian commitment to providing a phenomenology of culture, which accounts for the systematic unity and development of culture, and his Natorpian commitment to giving a bi-directional account of the subjective and objective dimensions of culture. I then turned to the complex framework of culture that Cassirer develops on this methodological basis. In order to explicate his framework of the unity of culture, I considered Cassirer's analysis of the basic free activity of spirit and categories of spirit that he sees as the conditions of possibility of culture as a whole. Next, I addressed his analysis of the subjective dimensions of culture, viz., the general and specific functions of consciousness that orient the lived-experience of individuals embedded in a cultural world. Turning to Cassirer's account of the objective dimensions of culture, I analyzed Cassirer's concept of a 'symbolic form' as a *forma formans* responsible for 'symbolically pregnant' configurations, that is, objects or practices, that make up our cultural world. I then surveyed Cassirer's system of symbolic forms. To this end, I took up the parallel Cassirer draws between the functions of consciousness and the symbolic forms. And I examined his defense of the irreducibility thesis, according to which each symbolic form is autonomous and cannot be reduced to any other, and the teleology thesis, according to which there is a teleological progression among symbolic forms toward the telos of culture, viz., consciousness of freedom. With this general framework for culture in place, we can now turn to the details of Cassirer's account of each individual symbolic form.

Notes

1 See Chapter Two for a discussion of Cassirer's criticisms of Cohen and Natorp for defending an overly rationalist account of culture.

2 Cassirer's interest in philosophical anthropology dates back to the late 1920s. The Davos disputation with Heidegger is oriented around the question, 'What is the Human Being?', and Cassirer titles his independent lecture, "Foundational Problems of Philosophical Anthropology." We also find evidence of Cassirer's interest in philosophical anthropology in the second chapter of his manuscript, "On the Metaphysics of Symbolic Forms," titled "The Problem of the Symbol as the Fundamental Problem of Philosophical Anthropology." See also ECN 6.

3 See *The Logic of the Cultural Sciences* (1942) for Cassirer's argument that the cultural sciences are as much as kind of science as the natural sciences (*Naturwissenschaften*), and his analysis of the difference between these sciences in terms of the phenomena and concepts ('law' and 'causality' versus 'style' and 'form') they organize themselves around (see LCS Studies 3–4).

4 Cassirer subtitles PSFv1 and v3 along these lines and he refers to v2 in these terms at PSFv2 16/16/13, 46/44/36, 96/90/76, 190/182/156, 201/197/168.

5 Translations of AH are my own.

6 For Cassirer's interpretation of the distinction between 'objective spirit' and 'absolute spirit' for Hegel, see *Das Erkenntnisproblem* Bd. 3 ECW 4, 313–314, 350–351; LCS 38/295; AH 59, 109.

7 Dilthey's influence on The Philosophy of Symbolic Forms is also evident in how
   Cassirer frames the distinction between 'cognition' in the natural sciences
   (Naturwissenschaften) and 'understanding' in the human sciences (Geisteswissenschaften)
   (PSFv1 lxxix/vii/69). According to Dilthey, the main methodological difference
   between these sciences turns on the fact that the natural sciences are oriented
   toward explanation (Erklärung) and cognition (Erkennen) and the human sciences
   are oriented towards understanding (Verstehen).
8 See, e.g., Cassirer's approval of Dilthey's project as one that is "not metaphysical,
   instead critical" (AH 110).
9 Natorp's formulation of the bi-directional approach has its roots in Kant's conception of the Transcendental Deduction as something that can be pursued either
   as a "subjective deduction" or "objective deduction" (CPR Axvii). See Cassirer's
   discussion of the subjective and objective deductions at PSFv3 8–12/8–12/7–
   11, 56–57/54–55/49–50; KLT 194/188.
10 Although Cassirer aligns the bi-directional approach with Natorp, insofar as
   Hegel's phenomenology of spirit involves analyzing both the 'subjective'
   dimension of culture in 'subjective spirit' and the 'objective' dimensions of culture in 'objective' and 'absolute' spirit, Hegel's is also a kind of bi-directional
   approach.
11 Cassirer also discusses the 'modality' or 'tonality' of the relations at PSFv2
   76–77/74–75/60–61, PSFv3 15/15/13.
12 It is important to distinguish the function of representation (Repräsentation)
   discussed in this section from what Cassirer elsewhere calls the 'function of
   presentation' (Darstellungsfunktion) and the 'function of pure signification' (reine
   Bedeutungsfunktion). The tendency to collapse the function of representation
   to the function of presentation is encouraged by the Manheim translation
   of Darstellungsfunktion as 'function of representation'; however, as I discuss, the
   function of presentation (Darstellungsfunktion) is a more specific function that
   orients consciousness in an objective direction and involves intuition, and should,
   as such, be distinguished from the function of representation (Repräsentation)
   as a general function that characterizes consciousness as a whole. And although
   Cassirer occasionally calls the general function of consciousness the 'function
   of signification' (Bedeutungsfunktion) (see, e.g., PSFv1 39/39–40/106), as I discuss, he standardly uses the 'function of signification' to refer to another
   more specific function of consciousness, viz., the function oriented to the
   "realm of pure signification [reinen Bedeutung]" (PSFv3 334/326/284, see also
   PS 262/261).
13 For Cassirer's argument that expressive perception is the key to solving the
   problem of other minds, see PSFv3 91–107/88–103/79–91.
14 Cassirer makes this point about language at LCW 338/115 and about technology at FT 276/142.
15 See, e.g., PSFv1 99/104/161, 113/119/174 and "The Kantian Elements of
   Wilhelm von Humboldt's Philosophy of Language" (1923).

16 Cassirer discusses the notion of symbolic pregnance at length in PSFv3: Ch. 5. See also ECN 4.
17 Cassirer connects *Prägnanz* in this sense to Leibniz's notion of '*praegnans futuri*', i.e., of "a future-filled and future-saturated now" (PSFv3 239/231/202).
18 In describing the relationship between symbolic forms in terms of 'harmony', Cassirer is not denying that there can be dramatic conflict between the forms. I return to this topic in Chapter Seven. See also LCS Study 5 for Cassirer's discussion of the 'drama' of culture.
19 See Friedman (2000): 101, 134–135; Gordon (2010): 248; Moss (2015): 247–250; Skidelsky (2008): 107.
20 See Krois (1987): 78–79, Lofts (2000): 34, Cornell and Panfilio (2010): 12, 93, Luft (2015): Pt. II, Ch. 3.4, Ch. 4.3.
21 See Chapter Eight for a discussion of this issue as raised by Blumenberg and Habermas. See Truwant (2014) for another discussion of how to reconcile the apparent tensions in Cassirer's view.

Further reading

Though many of these books also address the individual symbolic forms, see the following for a discussion of the general framework for Cassirer's philosophy of culture:

Friedman, Michael (2000). *A Parting of the Ways: Carnap, Cassirer, and Heidegger*. Chicago: Open Court. See pp. 98–110.
Krois, John Michael (1987). *Cassirer: Symbolic Forms and History*. New Haven: Yale University Press. See Chapters One and Two.
Lofts, S.G. (2000). *Ernst Cassirer: A "Repetition" of Modernity*. Albany: State University of New York Press. See Chapters One and Two.
Luft, Sebastian (2015). *The Space of Culture: Towards a Neo-Kantian Philosophy of Culture*. Oxford: Oxford University Press. See Part II.
Recki, Birgit (2003). *Kultur als Praxis: Eine Einführung in die Philosophie Ernst Cassirers*. Berlin: De Gruyter.
Skidelsky, Edward (2008). *Ernst Cassirer: The Last Philosopher of Culture*. Princeton: Princeton University Press. See Chapter Five.

See also:

Kreis, Guido (2010). *Cassirer und die Formen des Geistes*. Berlin: Suhrkamp. A systematic analysis of the philosophy of symbolic forms, which lays particular emphasis on the influence of Hegel and Dilthey.
Recki, Birgit, ed. (2012). *Philosophie der Kultur—Kultur des Philosophierens: Ernst Cassirer im 20. und 21. Jahrhundert*. Hamburg: Meiner. A comprehensive volume including essays on Cassirer's philosophy of culture in general, as well as his treatment of specific symbolic forms.

Truwant, Simon, ed. (2021). *Interpreting Cassirer: Critical Essays*. Cambridge: Cambridge University Press. See Part II for a discussion of Cassirer's philosophy of consciousness (Martina Plümacher, Guido Kreis) and Part III for a discussion of his philosophical methodology (Daniel Dahlstrom, Sebastian Luft) in the context of his philosophy of culture.

Verene, Donald (2011). *The Origins of the Philosophy of Symbolic Forms: Kant, Hegel, and Cassirer*. Evanston: Northwestern University Press. A discussion of the influence of both Kant and Hegel on *The Philosophy of Symbolic Forms*.

Six

The individual symbolic forms, Part I

From myth to natural science

Introduction

According to Cassirer, the symbolic forms are the "paths [*Wege*] by which spirit proceeds in its objectivization [*Objektivierung*]" (PSFv1 7/7/78). That is to say, the symbolic forms are the ways in which we, collectively, build up a cultural world around us, a world Cassirer calls the 'symbolic universe' or 'world of humanity'. And in the next two chapters, I explore Cassirer's account of the individual symbolic forms that make up our shared world.

In this chapter, my focus is on myth, religion, art, language, history, technology, mathematics, and natural science, and in the next on right (*Recht*). Although I have chosen these forms because they play an important role in Cassirer's philosophy of culture, this is not meant to be an exhaustive inventory. Cassirer conceives of the symbolic forms in open-ended terms, as something that historically unfold and develop through our ongoing efforts to build a cultural world. As such, any list of symbolic forms, including those that Cassirer offers, should be understood in provisional terms.[1]

In the following survey, I take my cue from the system that Cassirer develops in the third volume of *The Philosophy of Symbolic Forms*, which turns on a parallel between individual symbolic forms and specific functions of consciousness. As I discussed in Chapter Five, on Cassirer's view, individual symbolic forms objectively manifest the subjective, objective, and ideal orientations that are subjectively manifest in the specific functions of consciousness. Hence the following system of individual symbolic forms (see Table 6.1).

Table 6.1 The system of individual symbolic forms

Subjective: Specific function of consciousness	Objective: Symbolic form
function of expression (Ausdrucksfunktion)	myth, religion
function of presentation (Darstellungsfunktion)	art
	language, history, technology
function of pure signification (reine Bedeutungsfunktion)	mathematics, (mathematical) natural science, right (Recht)

One thing to note is the absence of philosophy in this system. However, according to Cassirer,

> It is characteristic of philosophical cognition as the "self-cognition of reason" that it does not create a principally new symbol[ic] form, it does not found in this sense a new creative modality—but it grasps the earlier modalities . . . as characteristic symbolic forms.
>
> ("Appendix: The Concept of the Symbol" 226/ECN 1, 264, translation modified)[2]

Instead of philosophy being its own symbolic form, Cassirer claims that philosophy is something that reflects on the other symbolic forms.[3] To be sure, Cassirer recognizes that not all philosophers have endorsed this conception of philosophy.[4] However, he thinks that this conception articulates what philosophy 'truly' is: "As long as philosophy still vies with these forms, as long as it still builds worlds next to and above them, it has not yet truly grasped itself" ("Appendix" 226/ECN 1, 264). Philosophy is thus absent from Cassirer's system of individual symbolic forms because it is something that stands in a reflective relationship to that system.

This caveat aside, in order to explore the above system of individual symbolic forms, I take my cue from Cassirer's irreducibility and teleology theses (see Chapter Five). According to the irreducibility thesis, the symbolic forms cannot be reduced to one another because they are autonomous from one another. With this thesis in mind, in what follows I consider the distinctive function of each symbolic form. To this end, I address the unique 'context of signification' and 'mode of objectivization' each symbolic form orients us toward in our efforts to build a cultural world.

Meanwhile, according to the teleology thesis, there is a teleological progression among the symbolic forms toward the telos of culture, viz., consciousness of freedom. More specifically, Cassirer argues that culture involves teleological progression towards the consciousness of the 'free activity of spirit' that underwrites our individual and cultural activities. Bearing this thesis in mind, below I examine Cassirer's account of the relationship of each symbolic form to the practical and theoretical consciousness of freedom.[5] With regard to the practical consciousness of freedom, I consider his account of the contribution each symbolic form makes to our recognition of our spontaneity and autonomy as agents. And with regard to the theoretical consciousness of freedom, I discuss the relationship each symbolic form has to our consciousness of the spontaneity and autonomy of spirit as that which the world conforms to. In this vein, I analyze the extent to which each symbolic form fosters our recognition that the basic structures of the world, viz., space, time, number, things, properties, and causes, do not belong to a substance absolutely independent from us, but are rather categories that have their source in the free activity of spirit.

To fill out the details, I start with an analysis of the symbolic forms he aligns with the function of expression, viz., myth and religion. Next, I turn to his account of art, as a symbolic form that blends together the tendencies involved in the functions of expression and presentation. I then address his account of the symbolic forms that align with the function of presentation alone, viz., language, history, and technology. Finally, I divide my analysis of the symbolic forms that parallel the function of pure signification between a discussion of mathematics and natural science in the last section of this chapter and of right in Chapter Seven.

Myth and religion

Although there are important differences between myth and religion, Cassirer aligns them both with the function of expression because they manifest the subjective and affective orientation of expression. As he puts it, in myth and religion,

> Whatever is seen or felt is surrounded by a special atmosphere— an atmosphere of joy or grief, of anguish, of excitement, of exultation or depression. Here we cannot speak of "things" as a dead or indifferent stuff. All objects are benignant or malignant, friendly or inimical, familiar or uncanny, alluring and fascinating or repellant and threatening.
>
> (EM 77/85)

Cassirer acknowledges that he is not alone in claiming that myth and religion have a subjective or affective basis; however, he takes issue with Freudian views, according to which this is a sign of the irrational or pathological basis of myth and religion (see, e.g., PSFv2 Chs. 1–2, MS Ch. 1). According to Cassirer, this subjective and affective orientation in myth and religion paves the way toward a distinctive kind of 'understanding' (PSFv1 lxxix/vii/69). And it is this 'understanding' that he seeks to elucidate.

Myth

According to Cassirer, the distinctive function of myth is to provide us with a feeling of everything, animate and inanimate, as belonging to a single whole. On his view, this is the context of signification that myth orients us toward and in which we produce mythical configurations, that is, objects and practices.

More specifically, he argues that the affective context of myth in terms of two dominant feelings: the feeling of the solidarity of life and the feeling of the sacred and profane. The former, Cassirer suggests, is the feeling that everything that exists, whether inanimate or animate, belongs to the same whole, viz., the whole of life (see, e.g., EM 82–86/90–93). Within this whole, Cassirer asserts that, "Life is not divided into classes and subclasses. It is felt as an unbroken continuous whole which does not admit of any clean-cut and trenchant distinctions" (EM 81/90). Moreover, he maintains that everything in this whole is fluid, being governed by the 'law of metamorphosis':

The limits between the different spheres are not insurmountable barriers; they are fluent and fluctuating. There is no specific difference between the various realms of life. Nothing has a definite, invariable, static shape. By a sudden metamorphosis everything may be turned into everything.

(EM 81/90)

On Cassirer's view, then, the feeling of the solidarity of life that pervades myth is the feeling that everything is part of the single, continuous, fluid whole of life.

In spite of this feeling of solidarity, however, Cassirer claims that there is still some differentiation in the mythical world and this is something that he attributes to the feeling of the sacred and the profane. He maintains that this feeling introduces two "different circles of significance [Bedeutungskreise]" into the whole of life: it demarcates the 'sacred', qua the 'uncommon' or 'transcendent', from the 'profane', qua the 'common' and 'everyday' (PSFv2 95–96/88–89/74–75).

On Cassirer's view, these two feelings provide the distinctive affective context of myth. It is in this context that we develop mythical 'configurations', including mythical objects and practices, like particular myths, totems, taboos, and rituals. And it is in this context and through these configurations that we develop the mythical understanding of the world around us and who we are.

If we now turn away from considerations about what is distinctive about myth toward teleological considerations, on Cassirer's view, the understanding of the world and who we are that myth encourages does not get us very far with respect to the theoretical or practical consciousness of freedom. To be sure, on Cassirer's view, our mythical world and mythical activities are, in fact, manifestations of our freedom, but he claims that the context and configurations of myth do not foster consciousness of this.

Beginning with the relationship between myth and the theoretical recognition of freedom, according to Cassirer, myth promotes a substance-based understanding of the world and of the categories of spirit. On Cassirer's view, myth encourages an understanding of the world as a real whole of life that is divided between the sacred and profane, which we are passive with respect to. From this perspective, space, time, numbers, things, properties, and causes are understood as parts of this substantial reality. For example, Cassirer claims that myth treats space as something that encompasses "two *precincts* [*Bezirke*] of being," viz., the sacred and profane, and a spatial direction, like east or west, as something that has "its own life" (PSFv2 107/100/85, 121/115/98, emphasis in original). So too in the case of cause, Cassirer suggests that myth presents impersonal forces, like mana or the orenda, and magical practices, like ritual or sacrifice, as a kind of real 'effective action' (*Wirken*) that brings about 'effects' (*Wirkungen*) (PSfv2 193/185/158–159). Likewise, with number, Cassirer maintains that myth identifies certain sacred numbers, like three or four, in substantial terms as having the power to take something profane and bring it into the domain of the sacred through a "process of 'hallowing' [*Heiligung*]" (PSFv2 170/170/143).

By Cassirer's lights, instead of fostering a recognition of space, time, number, things, properties, and causes as functions that have an ideal source in spirit, myth encourages us to regard them in substance-based terms as parts of a reality we are passive with respect to. Hence his claim,

> [the] *separation* of the ideal from the real . . . is alien to [myth] . . . Thus, accordingly we can almost call it a characteristic trait of mythical thinking that it lacks the category of the "ideal" and that, as a consequence, wherever it encounters something concerning pure significance [*rein Bedeutungsmäßiges*] . . ., in order to be grasped at all, [this significance] must be transcribed into something thing-like [*Dinglich*], into something being-like [*Seinsartig*].
> (PSFv2 48–49/47/38, translation modified)

For Cassirer, then, myth encourages an understanding of the world and of the categories in real, substance-based terms and thus occludes the fact that they are manifestations of our freedom.

Meanwhile, with respect to the question of the practical recognition of freedom, Cassirer clams that, on account of the real, substance-based orientation of myth, it promotes a more passive understanding of who we are (see, e.g., PSFv2, Pt. III). To this end, Cassirer suggests that myth promotes a substantial conception of who we are vis-à-vis a materialist conception of the soul,

> There is here still no "soul" as an independent unitary "substance" detached from the corporeal; rather, the soul is nothing other than life itself, to which the body is immanently and necessarily bound.
>
> (PSFv2 194/186/159)

The materialist conception is one Cassirer thinks is evident, for example, in the conception of the soul as continuing to take a corporeal form even after death, as a shade in Hades or as requiring material items, like coins (PSFv2 195–196/188–189/161–162).

Cassirer argues that, in myth, this materialist conception of the soul is coupled with a passive understanding of who we are as driven by forces outside of our control. These forces, Cassirer suggests, are sometimes conceived of as spiritual powers that exist outside us, which we are subject to, for example, in demonic possession (see PSFv2 192/184/158). At other times, however, Cassirer indicates that these forces are those of our own desires and needs.

This said, Cassirer acknowledges that myth encourages us to attempt to fulfill our desires by engaging in magical practices, rituals, and prayers; however, he suggests that this does not yet involve a robust conception of agency:

> at this stage there is as yet no true self. Through the magical omnipotence [*Allgewalt*] of the will, the I seeks to seize things and make them compliant; however, precisely in this attempt it shows itself still totally dominated, totally "possessed," by things.
>
> (PSFv2 192/184/158)

Cassirer's idea, then, is that insofar as the mythical subject remains passive with respect to the things that she needs and desires, she has not yet achieved an understanding of her 'true' self, that is, as an autonomous agent. And, as a result, Cassirer claims that myth only provides a 'simulacrum' of freedom:

> the I-feeling [in myth] . . . produce[s] only a simulacrum of effective action [Wirkens]. For all true freedom of effective action presupposes an inner bond [Bindung]. . . . The I comes to itself only in that it posits [setzen] these boundaries [Grenzen] for itself.
> (PSFv2 192–193/184/158, emphasis in original)

Thus, on Cassirer's view, since myth encourages us to regard the subject as someone who is determined by forces outside of her control, any sense of freedom we develop through mythical practices is an illusory one, falling far short of a practical recognition of our freedom as spontaneous and autonomous agents.

According to Cassirer, it is ultimately on account of this real, substance-based trend in myth that myth belongs on the lowest rungs of the ladder of culture. For, on his view, insofar as the affective context and configurations of myth encourage us to understand the world, categories of spirit, and who we are along substantial lines, myth does not put us in a position to be theoretically or practically conscious of our freedom.

Religion

On Cassirer's analysis, the relationship between myth and religion is a complicated one. On the one hand, he argues that myth and religion are intimately connected to one another:

> In the development of human culture we cannot fix a point where myth ends or religion begins. In the whole course of its history religion remains indissolubly connected and penetrated with mythical elements. On the other hand, myth . . . is from the beginning potential religion.
> (EM 87/96)

For this reason, Cassirer claims that many of the basic tendencies in myth, such as, the feeling of the "sympathy of the whole" and of the sacred and profane, shape religion too (EM 102/111). On the other hand, he insists that, as a distinct symbolic form, religion has a different function from myth. To this end, Cassirer argues that the function of myth is to disclose the world to us as divided between an ideal realm of the sacred that stands in opposition to the real realm of the profane, and to disclose who we are as morally responsible agents. It is in this context of signification that he takes religious objects and practices to emerge. And, on his view, it is on account of this function that religion advances our theoretical and practical consciousness of freedom beyond myth.

In more detail, Cassirer maintains that unlike in myth where everything is relegated to a single plane of reality, in religion the feeling of the sacred and profane undergoes a shift as the former comes to be identified with an ideal, spiritual world, and the latter with a real, sensuous world. As he puts it, in religion, the "character of holiness gradually transitions from individual persons or things [Sachen] to other . . . purely ideal determinations" (PSFv2 102/97/81). However, it is not just that religion introduces a distinction between the ideal/sacred and real/profane; he claims that it introduces the idea of the 'opposition' between them:

> The new ideality, the new spiritual "dimension," which is opened up [erschlossen] through religion, not only lends the mythical an altered "significance" [Bedeutung] but it literally introduces the opposition between "signification" and "existence" ["Dasein"] into the domain of myth. Religion takes the decisive step that is essentially alien to myth: in its use of sensible images and signs it at the same time knows them as such—as the means of expression that, though they reveal a determinate sense [Sinn], must necessarily at the same time remain inadequate to it.
>
> (PSFv2 290/280/239)

On Cassirer's view, religion takes a step beyond myth insofar as it introduces the opposition between what is real and sensuous, on the one hand, and what is spiritual, ideal, and sacred, on the other.

According to Cassirer, this distinction and opposition between the real and ideal provides us with an understanding of the world and the categories of spirit that takes us closer to a theoretical recognition of freedom. For, unlike in myth where space, time, number, things, properties, and causes are treated as parts of the real, substantial world, in religion we recognize that everything in the real sensuous world ultimately has its source in the ideal realm of the sacred.

Developing this point about time, for example, Cassirer argues that religion treats time as something that reflects the 'universal order' of the ideal world (see, e.g., PSFv2 136–141/131–137/112–116). In monotheistic religions, for example, he claims that time comes to be understood in terms of the ethical order that is ordained by the divine will: the future is conceived of as the ideal ethical end that the past and present are teleologically oriented towards (see, e.g., PSFv2 145/142/120, EM 99–100/109–110).[6] This ethical order thus becomes the true meaning of time and any temporal events that take place in the real world are understood as signs of this ideal temporal order. Cassirer thinks that a similar pattern can be traced with regard to the relations of space and number: their meaning is determined by the ideal order of the sacred and any spatial or numerical phenomena in the real world only have meaning in virtue of being signs of this ideal order.[7] In each of these cases, Cassirer thinks we see the trend in religion, which treats the world and categories of spirit as having an ideal source in the sacred.

This said, Cassirer argues that although religion encourages us to think of the categories of spirit in ideal terms, it operates with a conception of the ideal that has substance-based overtones. To this end, Cassirer claims that religion treats the ideal as something sacred that stands in opposition to the profane or sensuous:

> even the highest "truth" of religion remains attached to sensible existence. . . . It must continuously immerse and submerge itself in this existence whose ultimate "intelligible" purpose it strives to expel and eject from itself, because only in this existence does religious truth possess its form of manifestation [*Äußerungsform*] and hence its concrete reality and efficacy.
>
> (PSFv2 312–313/305/260)

Thus, instead of being understood in purely functional terms, Cassirer claims that in religion the ideal is understood in terms of a substantial struggle between the sacred and profane.

If we turn now to the practical dimensions of religion, according to Cassirer, religion provides us with an understanding of who we are as morally responsible individuals. Beginning with the point about individuality, according to Cassirer, unlike in myth where we understand ourselves as part of the 'whole of life', in religion we become increasingly aware of who we are as individuals. Cassirer traces this increase in our sense of individuality back to the development of increasingly personal conceptions of the divine. Indeed, Cassirer argues that different personalized conceptions of the divine promote different senses of individuality, and he traces this development through three stages (see EM 96–100/106–110).[8] In the first stage, religion develops 'special' or 'functional' gods, for example, those responsible for activities, like the harvest or hunting. And Cassirer claims that special gods promote a feeling for individuality as something defined in terms of the practical roles we play in society. The second stage involves the development of 'personal' gods, such as, the Greek gods, who are treated as quasi-persons endowed with "special mental gifts and tendencies" (EM 99/108). Cassirer maintains that this personal conception of the divine encourages a different kind of grasp of individuality, viz., one that defines individuality in terms of our talents and abilities (EM 99/108). Finally, in the third stage, there emerges an ethical conception the divine, for example, in monotheistic religions, like Zoroastrianism and the Prophetic religions, where the divine is defined in terms of "a great personal moral will" (EM 100/109). And, at this stage, Cassirer maintains that we come to develop an understanding of individuality, which turns on our sense of ourselves as morally responsible subjects.

Cassirer takes this last step to be particularly significant with respect to the question of our practical recognition of freedom because it involves "the expression of a new positive ideal of human freedom": it involves the recognition of human beings as morally responsible agents (EM 108/118). This said, although in religion the human being comes to see herself as a morally responsible agent, Cassirer emphasizes that there are still limits to this conception of freedom. Although religion encourages the subject to see herself as a morally responsible agent who is freely capable of choosing to act in accordance with laws, Cassirer claims that religion nevertheless continues to identify the divine as the source of these laws (see, e.g., MS 97–98/97–98, EM 100/109–110, AH 91–92). In this way, he maintains that religion ultimately promotes a conception of human subjects as dependent on the divine and sacred, which is outside of them. For this reason, Cassirer thinks that the practical recognition of freedom falls short of the full recognition of our spontaneity and autonomy.

Stepping back, on Cassirer's view, religion makes a distinctive contribution to culture by fostering an understanding of the world as divided between the ideal realm of the sacred and the real realm of the profane, and an understanding of who we are as morally responsible individuals. According to Cassirer religion advances us closer to the consciousness of freedom on both counts. In virtue of this understanding of the world, we come to regard space, time, number, things, properties, and causes not as substantial parts of reality that we are subject to, but as having an ideal source. And in virtue of this understanding of who we are, we come to develop a less passive and more active grasp of ourselves. Nevertheless, on Cassirer's view, in both cases, our theoretical and practical recognition of freedom is limited by a substantial conception of the ideal realm of the sacred as something we are passive with respect to.

Art

Although there is evidence that Cassirer hoped to dedicate a volume of The Philosophy of Symbolic Forms to art, this plan never came to fruition.[9] In order to understand his account of art as a symbolic form we must, instead, turn to his comprehensive treatment of it in Chapter Nine of An Essay on Man and in shorter essays and lectures, like "The Problem of the Symbol" (1927), "Mythical, Aesthetic, and Theoretical Space" (1931), "Language and Art" I and II (1942), and "The Educational Value of Art" (1943).

According to Cassirer, art is distinctive insofar as it occupies a place "between the world of pure expression and the world of pure presentation" (PS 268/267). More specifically, Cassirer argues that the function of art is to provide us with objective presentations of the subjective realm of expression. In defending this account of art, Cassirer distinguishes his view from 'expressivist' theories of art, according to which art is just a vehicle for the artist to express her emotions,[10] and mimetic theories of art, according to which art is just a mirror of nature. By Cassirer's lights, each of these theories veers too far toward the subjective or objective pole, neglecting the way in which these poles stand in "a purely correlative and mutually determining kind of relationship" in art (PS 268/268).

In his discussion of the expressive aspects of art, Cassirer does not restrict his focus to the way in which art expresses emotion or affect. Instead, on his view, the domain of expression encompasses everything that is 'intuitive', by which he means everything that we experience in a concrete, immediate, and sensuous fashion (EM 169–170/183–184). While emotions and affects certainly belong to this domain, Cassirer also claims that art can express the 'intuitive' forms of objects outside of us, which appear to us in a concrete, immediate, and sensuous way in space and time. For example, Cassirer argues that a landscape painter can express a landscape as we experience it, in its "individual and momentary physiognomy," which varies with "the play of light and shadow" and in "early twilight, in midday heat, or on a rainy or sunny day" (EM 168/182, 144/157). And on his view, part of what we do in art, is engage in a "process of concretion" in which we try and express the intuitive forms we experience in a concrete, immediate, and sensuous way, in as specific and individual a manner as possible (EM 143/155).

This said, Cassirer denies that art is just expression. Indeed, he claims that if an artist produces something in which she merely vents her emotions, this is 'sentimentalism', not art:

> The lyric poet is not just a man who indulges in displays of feeling. To be swayed by emotion alone is sentimentality, not art. An artist who is absorbed not in the contemplation and creation of forms but rather in his own pleasure or in his enjoyment of "the joy of grief" becomes a sentimentalist.
>
> (EM 142/155)

In order for something to count as art, Cassirer insists that the expressive tendencies must be coupled with presentative tendencies. Thus, on his view, art couples the process of expression and concretion with a "process of objectification" in which it presents what is experienced as concrete, sensuous, and immediate in an objective form (EM 142/155, 146/158).

In explicating the presentational aspects of art, Cassirer claims that artists deploy, what he calls 'sensuous forms', which are forms internal to an artistic medium, and which in turn present to us the intuitive forms of what we experience as concrete, immediate, and sensuous (see, e.g., EM 146/159, 154/167, 164/177, 168/182). When we encounter a work of art, we are thus encountering an objectification of the intuitive forms of, say, emotions or external objects, in the sensuous forms of an artistic medium. Cassirer claims that when we encounter intuitive forms in a work of art, we are able to engage with them through the more distanced attitude of 'contemplation' (EM 147/159, 165/178, 167/180). And, on Cassirer's view, through this contemplation, we can arrive at "knowledge of a peculiar and specific kind," viz., knowledge of these 'intuitive' forms (EM 169/183). Cassirer thus defends a kind of cognitivism, according to which art promises us knowledge of the intuitive forms we experience in an immediate, sensuous, and concrete way.

To illustrate, consider a Cassirer-style analysis of Audre Lorde's poem, "Coping" (1978). In this poem, Lorde expresses a feeling of coping. However, this expression is, at the same time, a presentation of this feeling: Lorde using poetic sensuous forms, like minimalistic language, undulating stanzas mirroring the image of small islands emerging from a sunless puddle, and the metaphor of a boy bailing out young seedlings in his flower patch, to provide us with an objective rendering of the experience of coping, which we can contemplate and learn from. Or to take a more externally-oriented example, from Cassirer's perspective, a painting like Georgia O'Keeffe's Black Mesa Landscape, New Mexico/Out Back of Marie's II (1930) expresses the intuitive form of the Black Mesa landscape, as it appeared to a viewer behind Marie's house around 1930. She makes the intuitive form of the landscape present to us by using various painterly sensuous forms, like a tripartite spatial organization, a contrast between warmer and cooler colors, and more attention to detail in the foreground. And by means of this painting, we can gain a kind of knowledge of that intuitive form.

For Cassirer, the symbolic form of art is distinctive insofar as it involves activities through which we express intuitive forms through the presentational means of art. And, on this view, these reciprocally related impulses toward expression and presentation thus provide the context of signification in which we produce artistic objects and develop artistic practices.

Let's turn now to the place of art within Cassirer's teleological account of culture. Beginning with the relationship between art and the theoretical recognition of freedom, according to Cassirer although art, like myth, orients us toward what is concrete, it provides us with a kind of distance that enables us to grasp the concrete in a freer light. And, as a result, art enables us to regard space, time, number, things, properties, and causes as concrete structures that are malleable and formable, rather than as substances we are wholly passive with respect to.

In "Mythic, Aesthetic, and Theoretical Space" (1931), for example, Cassirer argues that art operates with a 'concrete', rather than an 'abstract' conception of space and time, as something that we sensuously encounter, rather than something we just think about (MATS 328/422). And he claims that art gives us a more distanced relation to this concrete space and time, which allows us to grasp its structure. Making this point with respect to space, Cassirer claims, unlike our immersion in space in myth, art enable us to 'objectify' space and gain some 'distance' from it (MATS 328–329/423). In this more distanced attitude, Cassirer maintains that we achieve a "degree of freedom" in relation to concrete space and we come to grasp space as open to different "possible ways of configuration [Gestaltungsweisen]" through our 'formative' activities (MATS 328–329/422–423). So too in the case of time, Cassirer argues that art "makes [time] appear as if immersed in a particular 'color'" (MATS 331/425). For example, he suggests that epic poetry makes the 'color' of the 'past' appear, lyric poetry makes the 'color' of the present appear, and drama makes the 'color' of the future appear (MATS 331/425). In both of these cases, though art treats space and time as concrete structures, it provides us with a 'degree of freedom' in relation to those structures, which allow us to understand them in a more malleable light.

Extending this line of thought to other categories, on Cassirer's view, in the context of art, we come to grasp the concrete structures of space, time, number, things, properties, and causes with a 'degree of freedom' not achieved in myth. To be sure, insofar as these structures are still understood in concrete terms, art does not foster a fully theoretical recognition of them as categories of spirit. Nevertheless, it takes us toward a less substantial conception of them.

As for the practical recognition of freedom, according to Cassirer art is one of "the most powerful instruments of our inquiry into human nature" (EM 206/222). And though Cassirer thinks that there are a variety of ways in which art sheds light on human nature, here I shall focus on two ways that bear on the question of our practical consciousness of freedom.

In the first place, Cassirer claims art allows us to develop a more active relation to our affects and emotions, which helps us recognize that we are not wholly passive respect to them. According to Cassirer, art does this, in part by allowing us to adopt the more distanced attitude of contemplation in relation to emotions and affects that are presented in a work of art:

> The poet who represents a passion does not infect us with this passion. At a Shakespeare play we are not infected with the ambition of Macbeth, with the cruelty of Richard III, or with the jealousy of Othello. We are not at the mercy of these emotions; we look through them.
>
> (EM 147/160)

Indeed, for Cassirer, this is what the process of catharsis involves: "the soul acquires a new attitude towards its passions . . . [The artist] is not the slave but the master of his emotions; and he is able to transfer this mastery to the spectators" (EM 148/161). When this happens, Cassirer claims that our own affects undergo a 'transubstantiation': "in the realm of art even all our common feelings, our passions and emotions, undergo a fundamental change. Passivity is turned into activity; mere receptivity is changed into spontaneity" (EVA 211). In both of these ways, Cassirer argues that art advances our practical consciousness of freedom by revealing to us that we are not passive with respect to our affects and emotions.

Second, Cassirer argues that art also contributes to our consciousness of our freedom by providing us with the opportunity to exercise our freedom through aesthetic creativity. By Cassirer's lights, both the production and appreciation of a work of art requires creativity. According to Cassirer, artistic creation does not just involve the production of a piece, it also involves a creative mode of perception, which he calls 'perceptualization' (EM 151/163). As he describes it, perceptualization involves seeing not with "a passive eye that receives and registers the impressions of things," but instead with a "constructive eye" that fosters the "awareness of the pure forms of things," forms we often overlook in ordinary experience (EM 151/163, 144/156).

Meanwhile, on the side of the spectator, Cassirer claims that creativity is also required for the appreciation of art:

> Even the spectator of the work of art is not restricted to a mere passive role. In order to contemplate and to enjoy the work of art he has to create it in his measure. We cannot understand or feel a great work of art without, to a certain degree, repeating and reconstructing the creative process by which it has come into being.
>
> (EVA 212)

Here, Cassirer describes our appreciation of art in terms of 'contemplation' and 'enjoyment'. However, on his view this is not a passive state; it is a state in which we engage with a work of art in a creative and free fashion:

> In order to become aware of [aesthetic form], we have to produce it; and this production depends on a specific autonomous act of the human mind. We cannot speak of aesthetic form as a part or element of nature; it is a product of free activity.
>
> (EVA 211)

In underlining the connection between art and our creativity, Cassirer emphasizes the way in which the context of art provides us with a space in which we can become conscious of our freedom through its manifestation in our creative practices of production and creation.

Of course, on Cassirer's view, being conscious of our freedom with respect to our affects and emotions and our aesthetic creativity is not the same thing as being conscious of our full spontaneity and autonomy; however, art through its presentation of the 'world of expression' can enable us to take a step closer towards this recognition.

Language

According to Cassirer, language is like art to the extent that it manifests the objective tendencies operative in the function of presentation. However, Cassirer argues that the 'process of objectification' at work in language is fundamentally different from the one operative in art. Because art is governed as much by a subjective tendency as by an objective tendency, Cassirer claims that the objectification involved in art is oriented toward the concrete, immediate, and sensuous in all its "richness and variety" (EM 169/183). And though Cassirer by no means denies that we can use language as means of expression, he asserts that objectification is "the principle and most important task of human language" (EM 117/127, see also LCW 339/115–116, PSFv3 129–131/124–126/111–113). Cassirer thus insists that in its basic function language is oriented by the objective tendency involved in the function of presentation alone.

More specifically, Cassirer argues that language's primary function is to provide us with "our first objective or theoretical view of the world" (EM 135/146). By an 'objective' view of the world, Cassirer has in mind a view of the world that is populated by objects with properties, which remain constant, and which stand in stable spatial, temporal, and causal relations to one another. And he describes this as a 'theoretical' view of the world because this 'objective' world is one that we grasp in conceptual terms. So, unlike in art where the process of objectification is a 'process of concretion', in language Cassirer claims that this process amounts to a "process of abstraction," by means of which we develop a conceptual understanding of the objective world (EM 143/155).

In order to illustrate how language promotes objectification through abstraction, Cassirer points to our use of names. When a child learns to use a name, Cassirer argues that they "pass from a more subjective state to an objective state, from a merely emotional attitude to a theoretical attitude" (EM 132/142).[11] Cassirer takes this to be the case because when children learn to use names, they learn "to form concepts of . . . objects," and, by means of those concepts, they "come to terms with the objective world" (EM 132/143). For example, when a child learns a 'qualifying' name like 'red' or 'round' and 'classifying' names, like 'moon' or 'water', she forms concepts of properties and objects as things that remain constant over time (see PSFv1, Chs. 4–5). Through names, Cassirer claims that we thus acquire the ability to regard the world as a world of constant objects with constant properties, which we can grasp conceptually through thought. And in playing this role, Cassirer argues that names fulfill the primary function of language, viz., giving us an 'objective' and 'theoretical' view of the world.

According to Cassirer, though names illustrate these 'objective' and 'theoretical' tendencies in a particularly acute way, our linguistic 'configurations' as a whole are to be situated in this framework. Thus, whether we think of other types of words, like verbs or indexicals, sentences, grammatical practices, or entire languages, on Cassirer's view, they are oriented by this underlying function of language.

Turning to teleological considerations, Cassirer claims that the 'objective' and 'theoretical' tendencies of language advance our theoretical and practical recognition of freedom in a distinctive way. Beginning with the theoretical recognition of freedom, according to Cassirer, the theoretical tendency in language give us a more 'liberated' understanding of the world:

> language, with all its commitment to and entanglement with the world of the sensible and the world of the imaginative, reveals a tendency and force toward the logical-universal, through which it progressively liberates and attains to a purer and more independent spirituality of its form.
>
> (PSFv1 273/279/302)

For example, qualifying and classifying names enable us to form concepts of objects and properties, and indexicals and verbs enable us to detach from our 'entanglement' with the world in order to form concepts of space and time (see PSFv1 Ch. 3.1–2).

However, Cassirer argues that one of the main advances of language occurs when its 'process of abstraction' puts us in a position to develop a logical grasp of "pure form[s] of relation" (PSFv1 277/280/303).[12] In this vein, Cassirer singles out inflected languages as what most clearly promote a theoretical grasp of logical relations for, unlike in polysynthetic languages that use complex word-sentences, in inflected languages separate terms are used for nouns and verbs and this encourages a grasp of the relation as an independent unit (see, e.g., PSFv1 282–283/286–288/308–309). Among inflected languages, Cassirer suggests that those with a hypotactic structure, that is, those that use subordinating conjunctions and relative pronouns, that particularly enable us to form concepts of logical relations, through subordinating conjunctions, like 'because', 'therefore', and 'since' (PSFv1 285–286/292–293/310). And he claims that when languages shift away from using the copula 'is' in an attributive sense, that is, as qualifying some existing being, toward using it in a purely predicative sense, we develop a logical grasp of 'is' as a logical relation of 'synthesis' or 'combination' of concepts in a judgment (PSFv1 286/293/313).

Through the generalizing trend in language, Cassirer claims that it provides us with a more 'liberated' grasp of the categories, not just as substances that we are given, but as concepts and logical forms of relations. At the same time, however, Cassirer argues that this does not yet amount to a full theoretical recognition of freedom because there is a sense in which the generalizing tendencies in language always remain tethered to the sensuous world: "All linguistic presentation remains bound to the world of intuition and returns to it again and again. There are intuitive 'characteristic traits' that the process of linguistic naming singles out and holds fast to" (PSFv3 523–4/521/450, emphasis in original). Indeed, he takes this to be the case even at the most abstract level of achievement in language:

> Even where language progresses to its highest, specifically intellectual achievements, even where, instead of naming things or properties, occurrences or actions, it designates rather pure relations and relationships, this purely significative act does not, by and large, surpass certain limits of concrete-intuitive presentation. . . . Even the "is" of the predicative declarative sentence is for the most part designated linguistically in such a way that it preserves a secondary intuitive sense: the intellectual

relationship is replaced by a spatial one, a "being-here" [Da-sein]
or "being-over-there" [Dort-Sein]. . . . Thus, all logical determination
[Determination] belonging to language is originally contained in
its capacity and power of "demonstration."
(PSFv3 524/521–522/450, emphasis in original)

In its most advanced abstract phase, then, Cassirer insists that
language retains a 'demonstrative' character, as something that
remains oriented toward the sensuous world. Thus, even if language provides us with a more conceptual and logical grasp of the
categories of spirit than do myth, religion, or art, he claims that
this advance is nevertheless limited by language's tie to the sensuous world.

In addition to advancing our theoretical recognition of freedom,
Cassirer argues that language advances our practical recognition of
freedom.[13] To begin, Cassirer maintains that language encourages a
conception of who we are as subjects. Nouns and verbs, he suggests,
do this in an 'implicit' way, as they tacitly refer to a subject, whereas
pronouns do this in an 'explicit' way, by outright identifying a subject (PSFv1 210/225/259).

Furthermore, Cassirer contends that language gives rise to a more
robust sense of our volitional control. To this end, he argues that
language involves a 'retuning' of the volitional structure of myth:
whereas in myth, the subject is construed in passive terms as something that is overwhelmed by desires, needs, and emotions, language
gives the subject some control over them (LCW 346/275). He claims
that language does this, in part, by finding words for desires, needs,
and emotions that make them "'present and objective'" in a way that
undermines their ability to overwhelm the subject (LCW 350/279).
Moreover, he maintains that by providing the subject with a way
to articulate her desires, needs, and emotions, language becomes a
"medium of self-mastery" insofar as the verbal articulation of them
gives the subject some control over them (LCW 350/279). In these
ways, Cassirer suggests language enables us to gain some distance
from and control over our affects, emotions, needs, and desires and
so promotes a less passive and more active grasp of who we are.
This being said, as was the case with our theoretical recognition of
freedom, according to Cassirer the practical recognition of freedom

that we achieve in language remains limited because we continue to grasp who we are as tethered to the sensuous world.

In sum, on Cassirer's view, language is what gives us our first 'objective' and 'theoretical' grasp of the world. And although this more abstract grasp of the world advances our theoretical and practical consciousness of freedom, its progress is limited by its connection to the sensuous world.

History

As is the case with art, Cassirer does not dedicate a stand-alone volume of The Philosophy of Symbolic Forms to the topic of history; however, in An Essay on Man he explicitly labels history a 'symbolic form' and devotes a chapter to it. There, and in related texts like "The Philosophy of History" (1942),[14] Cassirer treats history as another symbolic form that is oriented in the objective direction of presentation. And he argues that the primary function of history is to provide us with a presentation of the past.

More specifically, Cassirer claims that history is responsible for presenting our 'life' in the past:

> the historian does not merely give us a series of events in a definite chronological order. For him these events are only the husk beneath which he looks for a human and cultural life—a life of actions and passions, of questions and answers, of tensions and solutions.
>
> (EM 187/201)

Insofar as history is oriented toward our 'life' in the past, Cassirer argues that it is 'anthropomorphic' in its aims; however, he describes this as an 'objective' kind of anthropomorphism:

> history is fundamentally anthropomorphic. It can live and breathe only in the human world. . . . But the anthropomorphism of historical thought is no limitation of or impediment to its objective truth. History is . . . a form of self-knowledge. . . . [T]he historical self is not a mere individual self. It is anthropomorphic

but it is not egocentric. . . . [H]istory strives after an "objective anthropomorphism."

(EM 191/206)

Thus, on Cassirer's view, the task of history is to present the human life from the past in an 'objective' form and to provide us with 'self-knowledge', that is, knowledge of human beings via the past.

In order to present the human life of the past in this objectively anthropomorphic way, Cassirer argues that historians rely on a method of 'interpretation' (see, e.g., EM 180–185/195–198). Since the past is past, Cassirer maintains that the historian can never have direct access to the human life she seeks to understand; instead, all she has at her disposal are cultural artifacts, documents, and other remnants from the past. Her task, then, is to read these as 'symbols' of life in the past (EM 175/189). As he sometimes makes this point, the historian must see in these configurations "characteristic facts," that is, facts that reveal the character of either an individual or an epoch (EM 196/212). For example, Cassirer claims that when the Cairo Codex, which contains fragments of comedies by Meander, was discovered in 1907, it was the historian's task to see it not just as a physical object, but to interpret it as a symbol of Greek life (see EM 175/189).

According to Cassirer, the process of interpretation ideally requires three things of the historian. First, a historian must engage in empirical investigation. In his words, the historian must "utilize all the methods of empirical investigation. He has to collect all the available evidence and to compare and criticize all his sources" (EM 204/220).

Second, in these empirical investigations, he maintains that the historian should be guided by sympathy: she must collect and assess the relevant facts in an unbiased and unprejudiced way. Cassirer points to Ranke as a model historian on this count:

> According to Ranke the historian is neither the prosecutor nor the counsel for the defendant. . . . He has to collect all the documents in the case in order to submit them to the highest court of law, to the history of the world. If he fails in this task,

if by party favoritism or hatred he suppresses or falsifies a single piece of testimony, then he neglects his supreme duty.

(EM 189/204)

Third, Cassirer maintains that the historian should make use of her imagination. This is because, on Cassirer's view, there is an element of creativity involved in historical interpretation. In lieu of a chronology, Cassirer claims that in order to present the life of the past the historian must offer an "*ideal reconstruction*" of it (EM 174/189, my emphasis). That is to say, the historian's interpretation is supposed to provide us with a 'resurrection' or 'rebirth' of the past and this is something she can accomplish only by reconstructing it in an imaginative way (EM 177/192). To be sure, Cassirer does not think she is thereby licensed to produce a "work of fiction"; she is 'bound' by the empirical sources and facts that she investigates (EM 204/220). Nevertheless, Cassirer maintains that her work should include an imaginative, indeed, dramatic element, "a great realistic drama, with all its tensions and conflicts, its greatness and misery, its hopes and illusions, its display of energy and passions" (EM 204/221).

Taking these three points together, on Cassirer's view, it is through the historical work of interpretation, which should involve a blend of empirical investigation, sympathy, and imaginative reconstruction that the symbolic form of history presents the life of the past in an objectively anthropomorphic way. And he conceives of this as the context of significance in which we engage in historical practices and produce particular histories.

With this basic picture of history in place, we can consider the teleological place of history in Cassirer's account of culture. With respect to the question of the theoretical recognition of freedom, Cassirer lays particular emphasis on the ways in which history provides us with an understanding of causality and time that is at once anthropomorphic and objective.

Regarding causality, Cassirer argues that history sheds light on a distinctive form of causality that he calls "individual causality" or "motivation," which pertains to the way in which the character of a person brings about certain events ("Philosophy of History" 128–129). For example, on Cassirer's analysis, the way in which Caesar's character leads him to cross the Rubicon is an instance of individual causality. According to Cassirer, it is important to recognize that this kind of causality is distinct from the sort of general causality at issue in natural science: whereas natural science is oriented towards ascertaining general causal laws, he claims that history should seek to understand individual causal motives (see EM 195–196/210–211). To be sure, he acknowledges that not all historians are so-motivated: he notes that Taine's naturalistic approach, Buckle's statistical approach, and Lamprecht's psychological approach all involve an analysis of the past in terms of general causal laws (see EM 195–201/ 210–217, LCS 77–86/435–445, PK Chs. XIV, XVII). Cassirer, moreover, does not deny that historical phenomena are shaped by various physical and psychological conditions. However, he thinks that it is a mistake for historians to offer a completely reductive analysis of the life of the past in terms of these general causal laws because such an approach neglects the central role that individual causality and motivation play in shaping this life.

Meanwhile, with respect to time, Cassirer argues that history gives rise to a new conception of time as "the time of human history" (MSF 91/89). Insofar as the time of human history is the time of the dynamic unfolding of human life, he claims time takes on a particularly dramatic character. To be sure, myth and religion also involve a dramatic conception of time; however, Cassirer takes the historical approach to involve a different conception of the relationship between past, present, and future. He maintains that whereas myth treats the past as something substantial that determines the present and future, in history the past loses its causal efficacy and becomes instead an object for reflection. When seen from this more distanced perspective, Cassirer argues that the past no longer appears as something that determines the present and future; it, instead, provides a lens by means of which we can reflect on them. To this end, he asserts that as an object of reflection, the past "gives us a freer survey of the present," helping us understanding how various 'conditions' and 'questions' from the past shape, though do not dictate, our present (EM 178–179/ 192–193). Meanwhile, with respect to the future, he suggests that the historical view of the past helps "strengthen our responsibility with regard to the future": though we acknowledge that the past is not something we are condemned to repeat, we are aware that it can shape the present in ways that we may need to struggle against (EM 179/193).

Like causality, then, Cassirer thinks that history discloses time to us from a perspective that is at once objective and anthropomorphic. And on his view, this advances our theoretical consciousness of freedom to the extent that it provides us with a human grasp of causality and time. That is to say, instead of regarding causality and time as substantial forces that are alien to the human being, history reveals them to us having a distinctively human texture. Insofar as this consciousness is mediated through the past, Cassirer thinks it is not yet a fully ideal grasp of causality and time, but it does take us closer to such a grasp.

Turning now to the question of history and the practical recognition of freedom, according to Cassirer, "Written and read in the right way history elevates us to an atmosphere of freedom amidst all the necessities of our physical, political, social, and economic life" (EM 206/222). On his view, history teaches us that we are not subject to these external necessities, but rather that we are the source of our own character, choices, and actions. Indeed, this is part of what the recognition of individual causality amounts to. Moreover, Cassirer claims that history reveals to us different ways in which other human beings have exercised their individual causality and the effects that can follow from certain characters, choices, and actions. This not only provides us with a sense of the real efficacy of our individual causality, but also provides us with models to reflect on with respect to the kind of agent we are or want to be. Finally, history can give us a sense of our collective agency and how certain large-scale patterns of character, choices, and actions manifest and take effect in the world. Though, on Cassirer's view, this practical recognition remains limited insofar as it is drawn from the past, history nevertheless advances our practical consciousness of freedom in these ways.

For Cassirer, then, the function of history is to provide us with a grasp of the past that is at once anthropomorphic and objective. And through this grasp, Cassirer claims that history advances our theoretical and practical consciousness of freedom, especially with respect to our understanding of our individual causality and the nature of time.

Technology

The final symbolic form that I shall discuss which Cassirer aligns with the function of presentation is technology. Though Cassirer lists technology as a symbolic form in the second volume of The Philosophy of Symbolic Forms and alludes to it in The Logic of Cultural Sciences and The Myth of the State, he offers his most extended analysis of technology as a symbolic form in "Form and Technology" (1930).[15] In treating technology as a symbolic form, Cassirer resists the tendency to understand technology exclusively in terms of technological products or achievements. Instead, he treats technology as a "forma formans," that is, an active "mode and type of production" by means of which we produce technological products (FT 276/142, 278/145). And as a symbolic form that parallels the function of presentation, Cassirer claims that the activity involved in technology is geared toward presenting something objective, viz., the world as something that can be shaped by means of our will.

In exploring how technology presents the world to us, Cassirer argues that, on the one hand, technology presents the world as a "world of 'objects'" that has a "foreign order," that is, an order that is independent from us (FT 290/157). Seen in this light, the world appears as a kind of "limit of the will, its counterpart and its resistance" (FT 290/157).

On the other hand, he maintains that in the context of technology we do not think of the world as 'fixed' and 'rigid', but rather as something 'formable' (FT 291/157–158). That is to say, although we recognize an objective order in the world, technology also enables us to see the world as something that we can form through our choices and actions. When we see the world as 'formable', Cassirer claims that we begin to see it as something that is governed by "rules of the 'possible'" (FT 291/158). And, a result, Cassirer asserts,

> technology repeatedly teaches us that the sphere of the "objective [Objektiven]" . . . never coincides with the sphere of that which is present [Vorhandenen]. . . . A technological creation . . . never binds itself to this pure facticity, to the given look of objects [Gegenstände]; rather, it obeys the law of pure anticipation, a forward-looking vision.
>
> (FT 309–310/176–177, translation modified)

According to Cassirer, in presenting the world to us as something that is at once objective and formable through our will, technology both advances our theoretical and practical recognition of freedom by means of particular conception of causality. With respect to the theoretical recognition of freedom, Cassirer argues that technology is what first gives rise to the concept of "objective causality," that is, to the concept of causality as something that does not just involve 'association', but rather "'necessary connection'" (FT 293–294/160–161). On his view, technology makes this possible because it provides us with the sense that there are fixed causal relations that connect together wills and objects in the world in a necessary way. And he claims that tools play a particularly significant role in this regard because they embody this "fixed rule" (FT 294/160). For example, a hammer embodies a necessary connection between a certain motion of a hammer and motion of a nail. In this way, Cassirer suggests that the intuition of tools paves the way for "the apprehension of 'objective causality'" in technology (FT 294/161). To be sure, on his view, this objective causality is not yet understood in fully ideal terms and so does not amount to a fully theoretical recognition of causality, but the grasp of necessary connection is a step in that direction.

Yet, according to Cassirer, technology does not just present the world of things as a world governed by an objective causal order; he argues it involves a practical grasp of causality that points toward the efficacy of our wills. In particular, he argues that over and above our awareness of the "power of mere desire," which overcomes us, technology gives us awareness of the "power of the will," as a power by means of which we can overcome the world, as it were, through technological means (FT 290/156). And insofar as technology presents those objects to us as ones that are independent from us, we come to grasp our will as a power that can 'master' what is independent from us (FT 296/163). In this way, Cassirer suggests technology "win[s] a new freedom," viz., the freedom of the will to form a world that is not only alien, but resistant to it (FT 296/163). And even though this conception of the will is indexed to the technological means through which we can achieve this freedom, on Cassirer's view, it does amount to a practical recognition of our volitional freedom.

In spite of these optimistic sounding claims, Cassirer is also sensitive to the worry that instead of increasing self-understanding, technology, in fact, leads to self-estrangement. According to Cassirer, although technology can, and has, led to this kind of estrangement, if we look at the underlying 'essence' or 'principle' of technology, we will find that this estrangement is not a necessary feature of it (FT 314/182). On his view, technology is not something that exists as "an end in itself," but rather is something that must be used in service of an end outside of itself (FT 315/182). And he maintains that much of the estrangement of modern technology is the result of technology being used in service of an "*economic system*," like capitalism, that alienates laborers from their work and encourages a consumerist culture (FT 314/182). However, from his point of view, if technology were, instead, used in service of an equitable system, then it would not necessarily lead to self-estrangement. To the contrary, he thinks that if used as a means to ethical ends, then technology could promote a more positive conception of labor, as involving "solidarity . . . in which all ultimately work for one and one works for all" and "'freedom through serviceability' [*Dienstbarkeit*]" (FT 315/183). Indeed, on Cassirer's view, promoting this sort of conception of labor, as something that reflects community, solidarity, and freedom is the "highest mission" of technology, which it could fulfill if only it were used for ethical purposes (FT 316/183).

For Cassirer, then, technology, with its revelation of the world as formable through our will, ultimately promises to advance consciousness of freedom specifically with respect to our theoretical and practical understanding of causality. In the former vein, technology provides us with a grasp of causality as involving necessary connection, and in the latter vein it provides us with a practical grasp of the power of our will. And though neither amounts to the full consciousness of our autonomy and spontaneity, it does advance our consciousness of freedom to a degree.

With this discussion of technology, I am at the end of my analysis of the symbolic forms that Cassirer situates on the lower and middle rungs of the 'ladder' of culture. Over the course of this analysis, we have seen the ways in which Cassirer thinks religion, art, language, history, and technology advance our theoretical and practical consciousness of freedom beyond myth. However, on Cassirer's view, none of these symbolic forms belongs on the highest rungs of the ladder of culture because they do not provide us with consciousness of our full spontaneity and autonomy. In order for this to happen, we need to arrive at a theoretical and practical recognition of freedom as something that has an ideal source, viz., in the 'free activity of spirit'. And it is only once we grasp this ideal, spiritual source that we will grasp space, time, number, things, properties, and causes as the functional categories of spirit that the world conforms to, and ourselves as spontaneous and autonomous agents. To reach this higher stage of consciousness of freedom, Cassirer claims we need to look to the symbolic forms that have an ideal orientation and align with the function of pure signification.

Mathematics and natural science

My analysis of Cassirer's account of the symbolic forms that parallel the function of pure signification will be divided between this and the following chapter. In what remains of this chapter, I return to Cassirer's account of mathematics and natural science, now situating them within his philosophy of culture. And in Chapter Seven, I discuss Cassirer's account of right (*Recht*). As we shall see, Cassirer groups these symbolic forms together as forms that have an ideal orientation and that provide us with "'objective' cognition" of the "realm of pure signification," that is, of pure concepts, laws, and principles (see, e.g., AH 74, PSFv3 334/326/284). The difference between them turns on the kind of cognition at issue: whereas mathematics and natural science give us "theoretical cognition" that bears on nature and "scientific experience," right gives us "practical cognition" that bears on the "world of willing and action [*Wollens und Tuns*]" and "social experience" (AH 71, 101, 75). And, according to Cassirer, the ideal concepts, laws, and principles that are uncovered in theoretical cognition are ones that govern the "realm of being" or what is, whereas those involved in practical cognition reflect the "realm of ought [*Sollen*]" or how we ought to will and act (DI 205/246).

These differences aside, Cassirer claims that these symbolic forms advance our consciousness of freedom more so than the other symbolic forms. He argues that mathematics and natural science advance our theoretical recognition of freedom, in particular, revealing to us the functional nature of theoretical laws, concepts, and principles that nature conforms to, and their ideal source in our spontaneity. Meanwhile, he maintains that right advances our practical recognition of freedom, fostering an individual and collective understanding of who we are as fully spontaneous and autonomous.

Setting the topic of right aside until the next chapter, I shall now focus on his account of mathematics and natural science. Though I discussed these topics in Chapters Three and Four, I have not yet considered them from the perspective of Cassirer's philosophy of culture. And when approached as symbolic forms, there are two further dimensions of mathematics and natural science that we need to take into account: their status as symbolic forms that manifest the ideal orientation of the function of pure signification and their teleological place in culture.

In order to bring out the significative tendencies of mathematics and natural science, it will be helpful to look at Cassirer's analysis of the relationship they have to expression and presentation. With regard to expression, Cassirer claims that the "most evident difference" about mathematics and natural science from the other symbolic forms is that "even in its earliest stage, [they have] fundamentally surpassed the world of mere 'expression'" (PSFv3 525/523/451). As he goes on to argue,

> The task of a *cognition* of nature already includes, with the imperfect means it has at its disposal, a conscious detachment from this world. "Nature," as an object of knowledge [*Wissens*], an object of thoughtful consideration and investigation, is given to the human being only once the human has learned to execute a cut between nature and its own world of "subjective" feeling.
>
> (PSFv3 525/523/451, emphasis in original)

He thus maintains mathematics and natural science orient us away from the world of expression entirely in our cognitive efforts.

According to Cassirer, the relationship that mathematics and natural science have to the function of presentation is more complicated.

On his view, in their initial stages, mathematics and natural science "cling to perception" and thus display tendencies at work in the function of presentation (PSFv3 525/523/451). However, if we look at the distinctive function of these symbolic forms, Cassirer asserts that they orient us toward breaking free from the limitations of perception and intuition:

> no mathematical theory of the natural event, is possible unless pure thinking detaches itself from the soil of intuition, unless it progresses to formations [*Gebilde*] that are of a fundamentally non-intuitive nature. And now the last decisive step is taken— now these very formations become the proper bearer of "objective" being. It is only through them that the lawfulness of being can be expressed: for they now constitute a new kind of object [*Objekt*] that may be designated as an object of a higher order over against those of the first phase.
>
> (PSFv3 374–375/367/320)

Cassirer claims that we see evidence of this function in the advance of mathematics and natural science towards a cognition of theoretical concepts, laws, and principles as functions that have an ideal provenance:

> Thought ultimately comes to a point, however, where this striving into the expanse of space and time no longer suffices, where, rather, a progress and transcending of a fundamentally different and more difficult kind is required of it. It must now not only tear itself away from the here and now, from the present place and moment [*Augenblick*], but must reach out beyond the whole of space and time, beyond the boundaries of intuitive presentation and of all presentability [*Darstellbarkeit*]. Thus, it must detach itself from the native soil . . . of intuition.
>
> (PSFv3 399/391/341)

By Cassirer's lights, this trajectory in mathematics and natural science reveals to us that, at the most fundamental level, these symbolic forms have a primary function that is ideal in orientation.

Cassirer, furthermore, takes the ideal orientation of mathematics and natural science to have teleological implications insofar as they advance our theoretical consciousness of freedom. According to Cassirer, mathematics and natural science are the first symbolic forms to grasp the categories of spirit as functions that have their source in the activity of spirit and as that which nature conforms to.

The functional nature of Cassirer's philosophy of mathematics and natural science is something I discussed in Chapters Three and Four. On his view, mathematics and natural science progress toward a purely functional grasp of space, time, number, thing, and cause. As we saw, Cassirer claims that Dedekind's arithmetic, Poncelet's projective geometry, Klein's Erlangen Program, and Hilbert's axiomatics should be understood in terms of a development away from a substance-based view and toward a function-based view of numerical and spatial concepts. Similarly, in modern physics and chemistry, Cassirer maintains that we find a shift away from a substance-based view of 'things' and 'causality' towards a function-based view. With regard to the things, for example, he maintains that natural science treats them not as what is given in intuition, but rather as 'physical objects' that results from the transformation of an empirical manifold in light of the functional concepts, laws, and principles of mathematics and natural science.[16] Meanwhile, concerning causality, he argues that in physics causality is ultimately understood in categorial terms as the 'functional dependence of magnitudes' and 'conformity according to law'. With each of these categories, Cassirer thus argues that mathematics and natural science advance us toward a purely functional conception of them, as ideal functions or relations that nature conforms to.

However, on Cassirer's view, mathematics and natural science also provide us with an ideal grasp of these functions as ones that have their source in the free activity of spirit. As Cassirer makes this point, mathematics and natural science involve a "specific mode of 'reflection'" on the thought of spirit (PSFv3 339/331/288). And in this reflection, Cassirer claims that the categories reveal themselves to be "*created* by thought itself" and produced "in full freedom, in pure self-activity [*Selbsttätigkeit*]" of spirit (PSFv3 335/327/285, emphasis in original). Thus, on Cassirer's view, mathematics and natural science ultimately give us a grasp of the categories of spirit, as such. And it is for this reason that Cassirer maintains that with mathematics and natural science, "the movement of spirit is not obstructed and destroyed . . . but rather that spirit for the first time truly *discovers itself* as that which contains within itself the principle, the beginning of movement" (PSFv3 400/392/341, my emphasis). As we see here, on Cassirer's view, mathematics and natural science ultimately advance the theoretical consciousness of freedom to a higher level insofar as they give rise to consciousness of the categories as functions that have an ideal source in the free activity of spirit and which nature conforms to. And it is in this way that Cassirer argues that these symbolic forms advance culture farther than do myth, religion, art, language, history, or technology.

Though this theoretical recognition of freedom gives us a certain kind of self-understanding, Cassirer claims that mathematics and natural science do not provide us with a practical grasp of our spontaneity and autonomy as agents. As he makes this point in *Substance and Function*, "The 'individual' of natural science includes and exhausts neither the individual of aesthetic consideration nor the ethical personalities" (SF 232/254). Addressing this topic at more length in the final chapter of *Determinism and Indeterminism*, Cassirer argues that the conclusions from quantum mechanics do not bear on morality because the 'order' we uncover through the former is not a moral 'order':

> [the moral problem of freedom] cannot be solved by simple reduction to natural laws but has to be based on an independent type of orderliness according to law [*Typus von Gesetzlichkeit*], the autonomous orderly structure of the will. If this is kept in mind, it becomes understandable why ethics has nothing to fear and little to hope from the changes in the basic concepts of science which have taken place in modern physics.
>
> (DI 206/248, emphasis in original)

From Cassirer's perspective, we must look outside of mathematics and natural science for cognition of the distinctive kind of 'orderliness according to law' that governs an autonomous will and advances the consciousness of freedom along practical lines. Indeed, as we shall see in the next chapter, on Cassirer's view, this task ultimately falls to the symbolic form of right.

Summary

Having established the general structure of Cassirer's philosophy of culture in the last chapter, my goal in this chapter was to begin teasing out the details of Cassirer's analysis of the individual symbolic forms. Throughout this survey, I endeavored to highlight both the distinctive contribution that each symbolic form makes to culture, as well as the teleological progression Cassirer takes there to be among the symbolic forms as they advance our consciousness of freedom. To this end, I considered how each symbolic form contributes to our theoretical recognition of our freedom vis-à-vis a grasp of the categories of spirit as functions that have an ideal source in spirit and that the world conforms to. Second, I examined how each symbolic form contributes to our practical recognition of our freedom as autonomous and spontaneous agents. And I organized this discussion around Cassirer's alignment of individual symbolic forms with the specific functions of consciousness.

I opened with Cassirer's account of myth and religion, as symbolic forms that align with the function of expression and are thus subjective and affective in orientation. I examined his account of myth as something that provides us with a feeling of everything belonging to the whole of life, which is divided into the sacred and profane. And, from the teleological perspective, I discussed his account of the limitations which the real, substance-based outlook of myth places on our theoretical and practical recognition of freedom. I then addressed his account of religion, as something the gives us a grasp of the world as divided between an ideal realm of the sacred, which is opposed to a real realm of the profane. We saw that, on his view, religion advances our theoretical consciousness of freedom toward an ideal grasp of the categories of spirit and the practical consciousness of freedom toward a grasp of who we are as practical agents; however, we also saw he regards these achievements as limited by a conception of the ideal that is wedded to the opposition between the sacred and profane. Turning to art,

I examined Cassirer's treatment of it as something that enables us to objectively present the world of expression and the intuitive forms of what we experience in a concrete, sensuous, and immediate way. And I analyzed his claim that art teleologically advances our theoretical grasp of the categories by revealing them as concrete structures that are malleable, and our practical grasp of who we are as capable of 'mastering' our emotions and creatively exercising our freedom.

Next, I analyzed his account of language as a symbolic form whose primary function is to present us with an 'objective' and 'theoretical' view of the world. I highlighted Cassirer's treatment of how language advances our theoretical recognition of freedom through this conceptual and logical grasp of the categories, and our practical recognition of freedom as subjects who can exercise control over our emotions, needs, and desires. I then explicated Cassirer's account of history, as a symbolic form whose function is to give us an objective anthropomorphic grasp of the human life of the past. And I considered the ways in which Cassirer thinks history can advance our theoretical grasp of causality and time and our practical grasp of our individual and collective causal efficacy. Turning to technology, I looked at Cassirer's argument that technology is a symbolic form that presents us with the objective world as something that we can form through our wills. Then I examined Cassirer's account of how this advances our theoretical understanding of objective causality as something that involves necessary connection and our practical understanding of the power of our wills. At the end of the chapter, I started exploring his account of the symbolic forms that parallel the function of pure signification, with a discussion of mathematics and natural science, as symbolic forms that give us theoretical cognition of the ideal concepts, laws, and principles that govern nature. And I looked at his account of how this cognition advances our theoretical consciousness of freedom to a high degree, by revealing the categories as functions that have an ideal source in the spontaneity of spirit and as that which the world conforms to.

I shall continue pursuing this survey in the next chapter, as I take up Cassirer's account of right and the way it advances our practical consciousness of freedom. However, as we saw in Chapter Five, although Cassirer thinks that culture is capable of progress in these ways, he does not conceive of progress as a guarantee. And as I now take up the moral and political dimensions of his philosophy of culture, it will become clear that there, more than anywhere else, Cassirer calls our attention to the risk and responsibility surrounding the progress of culture.

Notes

1 Cassirer's list at PSFv2 xxx/xi/xiv includes myth, art, (mathematical-scientific) cognition, ethics, right, language, and technology; at FT 275/142 includes economics, the state, morality, right, art, religion, and technology; at LM 168/266 includes language, cognition, art, right, ethics, basic forms of the community, the state, myth, and religion (he cites this list again at AH 82, fn. 1). For a discussion of the question of the number of symbolic forms, see Luft (2015): Ch. 4, Section 2. For a discussion of comedy as a symbolic form, see Marra (2015).

2 See also Cassirer's description of philosophy in CP (1935).

3 For an alternative reading of this passage as allowing for philosophy to count as a symbolic form, see Capeillères (2007).

4 This is a conception of philosophy that Cassirer traces back not just to Kant and Hegel, but to Renaissance philosophy: "Since the days of the Renaissance, philosophy has brought all the powers of modern thought before its forum, questioning them about their meaning and right, their origin and validity" (FT 274/141).

5 I thus set aside questions about the teleological progress internal to each symbolic form as it progresses through the mimetic, analogical, and symbolic phases.

6 Cassirer's analysis of the temporal dimensions of religion was influenced by Cohen's treatment of religion, e.g., in The Concept of Religion in the System of Philosophy (1915) and Religion of Reason Out of the Sources of Judaism (1919).

7 Cassirer explores the topics of the development of space, time, and number in myth and religion in PSFv2, Pt. 2, Ch. 2.

8 In tracing this progression of the conception of the gods, Cassirer follows Hermann Usener, see, e.g., LM §2.

9 In a letter to Paul Schlipp about An Essay on Man, Cassirer says, "I would give here for the first time a thorough presentation of my theory of aesthetics. Already in the first sketch of The Philosophy of Symbolic Forms a particular volume on art was considered but the disfavor of the times postponed its working out again and again" (quoted in SMC 25).

10 Cassirer associates this kind of expressivist theory with Benedetto Croce and R.G. Collingwood.

11 In addition to discussing this point in relation to children's use of names (e.g., EM 132/144, PSFv3 140–141/134/121–122, LCW 341/117–118), Cassirer

makes it with reference to Laura Bridgeman and Helen Keller's use of names (e.g., PSFv3 131/125/112; EM 33–38/38–43, 131–133/142–143) and to research on aphasia and psychic blindness (e.g., PSFv3, Pt. 2, Ch. 6.2).

12 This is the topic of PSFv1: Ch. 5. For the lists of these relations, see PSFv1 284/289/310, 286/292/313.

13 This is the topic of PSFv1 Ch. 3, Section 4.

14 He also lists history alongside the other symbolic forms in MSF 85–92/83–90. See also his historical analysis of the development of history in the Enlightenment in *Philosophy of the Enlightenment* Ch. 1 and "Descartes, Leibniz, and Vico" (1941–42) and of historicism in the 19th century in PK: Part III and "Hegel's Theory of the State" (1942).

15 Cassirer lists technology as a symbolic form at PSFv2 xxx/xi/xiv. He also discusses it at PSFv2 252–256/250–253/212–216, MSF 41/39–40, LCS 26–27/382–384 and in his discussion of the 'technique of myth' in MS, which I take up in Chapter Seven.

16 For Cassirer's discussion of the difference between a 'thing' and 'object', see PSFv3, Pt. 3, Ch. 2.

Further reading

In addition to suggestions at the end of Chapter Five, see the following.

Biagioli, Francesca (2015). "Cassirer's View of the Mathematical Method as a Paradigm of Symbolic Thinking." *Lectiones & Acroases Philosophicae* 8(1): 193–223. A discussion of the relationship between Cassirer's philosophy of mathematics and the philosophy of symbolic forms.

Bundgaard, Peer (2011). "The Grammar of Aesthetic Intuition: On Ernst Cassirer's Concept of Symbolic Form in Visual Arts." *Synthese* 179: 43–57. A discussion of art as a symbolic form.

Capeillères, Fabien (2008). "History and Philosophy in Ernst Cassirer's System of Symbolic Forms" in *The Symbolic Construction of Reality: The Legacy of Ernst Cassirer*, edited by Jeffrey Andrew Barash. A discussion of history as a symbolic form.

———(2012/13). "Artistic Intuition Within Cassirer's System of Symbolic Forms. A Brief Reassessment." *Cassirer Studies* V/VI, 91–124. A discussion of art as a symbolic form.

Heis, Jeremy (2015). "Arithmetic and Number in the Philosophy of Symbolic Forms." In *The Philosophy of Ernst Cassirer: A Novel Assessment*, edited by J. Tyler Friedman and Sebastian Luft, 123–140. Berlin: De Gruyter. A reading of the ways in which Cassirer revises his theory of mathematics in *The Philosophy of Symbolic Forms*.

Hoel, Aud Sissel and Folkvord, Ingvild, eds. (2012). *Ernst Cassirer on Form and Technology: Contemporary Readings*. Houndmills: Palgrave Macmillan. A volume of collected essays dedicated to Cassirer's "Form and Technology."

Langer, Susanne K. (1949). "On Cassirer's Theory of Language and Myth." In *The Library of Living Philosophers: The Philosophy of Ernst Cassirer*, edited by Paul Arthur Schlipp, 379–400. Evanston: Library of Living Philosophers. A discussion of language and myth as symbolic forms.

Moss, Gregory (2015). *Ernst Cassirer and the Autonomy of Language*. Lanham: Lexington Books. A discussion of language as a symbolic form.

Pollok, Anne (2015). "The First and Second Person Perspective in History: Or, Why History is 'Culture Fiction'." In *The Philosophy of Ernst Cassirer: A Novel Assessment*, edited by J. Tyler Friedman and Sebastian Luft, 341–360. Berlin: De Gruyter. A discussion of history as a symbolic form.

Schmitz-Rigal, Christiane (2011). "Science and Art: Physics as a Symbolic Formation." *Synthese* 179: 21–41. A discussion of physics as a symbolic form.

Truwant, Simon, ed. (2021). *Interpreting Cassirer: Critical Essays*. Cambridge: Cambridge University Press. A collected volume dedicated to Cassirer's philosophy of culture. See Part I for essays on language (Robert Leib), art (Samantha Matherne), history (Anne Pollok), science (Massimo Ferrari), quantum mechanics (Thomas Ryckman), technology (Nicolas de Warren), and politics (Simon Truwant).

Seven

The individual symbolic forms, Part II

The ethics and politics of culture

Introduction

During Cassirer's student years at Marburg, he received the nickname 'the Olympian', having the reputation of being removed from everyday matters, absorbed in thought instead. This perception of Cassirer is one that not only followed him throughout his career, but also colors his reception still today, as his philosophy is often characterized as the theoretical work of an erudite thinker for whom practical issues, like morality and politics, never play a central role. To be sure, in his last book, The Myth of the State (1946), Cassirer addresses the modern political situation and the rise of fascism; however, even with this text, many readers feel that he waded into the political scene with reluctance and without a substantial moral and political philosophy to support his views. We see this appraisal, for example, in Leo Strauss's review of The Myth of the State, where he argues that what we get from Cassirer are 'inconclusive' remarks on the relationship between myth and politics, when what was really needed was "a radical transformation of the philosophy of symbolic forms into a teaching whose center is moral philosophy" (Strauss 1947: 128).[1] Though this reading of Cassirer is not uncommon, it fails to do justice to the ethical and political dimensions of his philosophy of culture.[2]

There is one practical dimension of Cassirer's philosophy of culture that I have already considered, viz., his account of the progress of culture as progress toward freedom. Cassirer, indeed, identifies the "problem of freedom" as the "ethical problem of culture" (CIPC 82/112). He thus conceives of the teleological progress that he details in The Philosophy of Symbolic Forms as an ethical matter. However, in the 1930s and 1940s Cassirer's ethical and political views develop in two significant ways. First, he provides an account of morality and right (Recht) within the framework of the philosophy of symbolic forms. Second, he offers a diagnosis of the rise of fascism in terms of a regression towards a more mythical outlook. Although in pursuing this program Cassirer draws on the model of liberalism he defended in Freedom and Form (1916), he develops it in new directions in texts including "The Idea of the Republican Constitution" (1928),[3] "Transformations of the Ethos and Theory of the State in German History" ("Wandlungen der Staatsgesinnung und der Staatstheorie in der deutschen Geschichte") (1930), "On the Nature and Development of Natural Right" ("Vom Wesen und Werden des Naturrechts") (1932), Ch. 13 of Determinism and Indeterminism: "Concluding Remarks and Implications for Ethics" (1936), "Critical Idealism as Philosophy of Culture" (1936), Axel Hägerström (1939), "Judaism and the Modern Political Myths" (1944), and The Myth of the State (1946).

My aim in this chapter is to explore these moral and political aspects of Cassirer's philosophy of culture. I begin with a discussion of the account of morality and right that he develops. I take my cue from Axel Hägerström, which is, arguably, Cassirer's most extended analysis of morality and right within the framework of the philosophy of symbolic forms. I explore two aspects of his account in Axel Hägerström. First, I examine his analysis of 'ethical consciousness', as something that is cognitive in nature and capable of objectivity. In this section, I also discuss his account of the 'phenomenology of ethical consciousness', which he defines as the progressive development of moral theories. Second, I look at his account of right as a symbolic form. In this discussion, I address his account of natural right (Naturrecht), the social contract, and the idea of a republican constitution. Then, I turn to Cassirer's diagnosis of the rise of fascism in the 20th century as the result of a regression in culture towards a more mythical outlook. I consider his account of how the fascist 'technique of myth' precipitated this mythical outlook, as well as his discussion of the indirect culpability of some philosophers.

I conclude with an analysis of the moral duties and responsibilities to culture that Cassirer thinks this political crisis brings to light.

Morality and right

Though in The Philosophy of Symbolic Forms Cassirer by no means presents his philosophy of culture in exclusively theoretical terms,[4] in Axel Hägerström, he sets himself the task of filling out the practical dimensions of his view,

> I have used the stimulus, which my study of Hägerström's major works has provided, to sharpen my own fundamental view [Grundanschauung], which I presented particularly in my Philosophy of Symbolic Forms, and to apply it to a new area. Thus my overall conception [Gesamtauffassung] of the problems of ethical philosophy and philosophy of right is treated much more extensively here than in my earlier writings, which especially concerned theoretical philosophy.
>
> (AH 7)[5]

Arguably, Axel Hägerström thus represents Cassirer's most sustained attempt to work out an account of morality and right from the perspective of the philosophy of symbolic forms.

Axel Hägerström is, of course, also a study of the philosophy of Axel Hägerström, whose work Cassirer became familiar with while living in exile in Sweden. Hägerström was a leading figure of the positivist movement, 'Scandinavian Realism', and he was guided by the motto, "Ceterum censeo metaphysicam esse delendam" ("furthermore I propose that metaphysics must be destroyed") (AH 58).[6] He is perhaps best known now for his legal realism. Skeptical of traditional theories of natural right (Naturrecht), Hägerström defends a version of legal realism, according to which legal concepts should be understood on the basis of reality, not on the basis of anything metaphysical or ideal. However, Hägerström's legal realism is part of an overarching philosophical view, which includes a theoretical and practical component. In his theoretical philosophy, Hägerström argues that the cognition in mathematics and natural science is not grounded in anything metaphysical, but rather in reality. And in a non-cognitivist vein in his moral philosophy, Hägerström defends an emotivist theory of morality, according to which moral judgments are merely subjective expressions of emotion. In light of this emotivism, he endorses a kind of moral relativism, which treats all moral judgments as equivalent qua emotive expressions. And, according to Hägerström whatever 'necessity' or 'must' we feel in the ethical domain is ultimately the result of mechanical processes of association (AH 61).

In *Axel Hägerström*, Cassirer thus presents his own account of morality and right in dialogue with the theoretical, moral, and legal aspects of Hägerström's philosophy. For my purposes, the most significant chapters are Chapter Three, "The Moral Philosophy," and Chapter Four, "Myth and Right." In these chapters, Cassirer agrees with Hägerström's rejection of a metaphysical approach to morality and right. However, instead of endorsing a positivist approach, Cassirer endorses a critical approach in which he analyzes morality and right in terms of 'functions' that serve as conditions of the possibility of the 'facts' of the "world of willing and action" (AH 74–75, 79, 86–87, 94, 101–102). To this end, he develops an account of ethical consciousness, on the one hand, and of right as a symbolic form, on the other.

Ethical consciousness

Cassirer orients his discussion of ethical consciousness around the "*factual* conduct [*faktische Verhalten*]" of this consciousness (AH 63, emphasis in original). He focuses, in particular, on the ethical consciousness involved in "the lived-experience of 'evaluation' or 'opinion' [*das Erlebnis der 'Bewertung' oder der 'Stellungnahme'*]" in ethical matters (AH 62). And in developing this account of ethical consciousness, Cassirer takes issue with Hägerström's non-cognitivist, relativist, empirical account of it.

Beginning with his rejection of Hägerström's non-cognitivism, according to Cassirer, if we look at the 'lived-experience' of ethical consciousness, we find that it is something that is cognitive in structure (AH 62). More specifically, Cassirer argues that in ethical consciousness we do not experience ourselves as overwhelmed by affects and emotions; we experience ourselves as having a

'personality' or 'character', which is guided by certain normative principles for how we should will and act (AH 62, 64). Thus,
when we make an ethical evaluation or form an ethical opinion,
Cassirer claims that this has a cognitive structure, viz., we 'subsume' a particular case under a more general principle (AH 63).
For example, suppose I form an ethical evaluation that it would be
good to keep a promise to a friend. On Cassirer's view, this evaluation does not just involve a positive feeling; it involves me making
a judgment in which I subsume this particular case under a general principle, 'I should keep my promises'. This said, according
to Cassirer, the particular normative principles that we endorse
will vary given the particular 'personality' and 'character' that we
have, for example, our principles could be tied to a religious or
moral theory. Nevertheless, on his view, our moral evaluations and
opinions reveal themselves to have a cognitive structure in which
our normative principles serve as a 'premise' from which certain
'conclusions' follow (AH 64).

Cassirer, moreover, argues that in our 'factual' ethical consciousness, we are sensitive to a certain 'regulative' 'demand' (Forderung)
that is placed on us: to commit ourselves to principles that are 'universal' in character, that is, applicable in all circumstances and to
all persons (AH 76). Cassirer's idea seems to be that insofar as we
have a 'personality' or 'character', we strive to bring maximal 'unity'
and 'inner consistency' into our choices and actions (AH 64). He
describes this goal in terms of the "idea of 'unity of willing'"
(AH 75). And he claims that a universal principle represents the
sort of principle that would bring maximal unity and consistency
into our volitional lives: it is a principle that we can act on in all
circumstances, and since it is a principle that governs all wills, it
will make my willing consistent with how one ought to will in general. Thus, on his view, a universal principle is one that enables us to
most closely approximate the idea of 'unity of willing'. For Cassirer,
then, it is built into our ethical consciousness to strive toward universal principles as the key to the unity of our character and volitional lives.

However, by Cassirer's lights, once we appreciate the role that
'unity' and 'universality' play in ethical consciousness, we are in a
position to see why Hägerström's relativism is misguided. According

to Cassirer, 'unity' and 'universality' serve as the 'standard' (*Maßstab*) by means of which we can measure the 'objectivity' of our moral judgments and moral theories (AH 69). Cassirer's basic idea is that the objectivity of a moral judgment or theory depends on how universal its principle is, that is, how much of the domain of willing and action it normatively governs.

In order to clarify this standard, Cassirer offers an extended analysis of the parallels between objectivity in the moral domain and in the theoretical domain, specifically in the domain of natural science. As I discussed in Chapter Four, on Cassirer's view, it is a mistake to conceive of 'objectivity' in natural science in a substance-based way, that is, in terms of the 'correspondence' between our theories and an absolutely mind-independent reality (AH 69). Instead, he endorses a function-based view, according to which objectivity is measured in terms of the regulative demand of 'unity' and 'universality' (see AH 66–74). That is to say, one natural scientific theory is more objective than another if it accounts for relations that are invariant with respect to a wider sphere of scientific experience and physical phenomena.

Similarly, Cassirer argues that "unity and universality" serves as the "standard" of "'objective cognition'" in the practical domain (AH 74). He treats this as a standard that measures the objectivity of both moral judgments made by individuals and moral theories: one moral judgment or moral theory will be more objective than another depending on whether its guiding normative principle is one that can govern all volitional experience and physical phenomena in a systematic way. A judgment or theory that reflects a normative principle that is valid only for a certain set of people or set of circumstances, for example, a principle tied to customs or a particular religious doctrine, will be less objective than a normative principle that is valid for all people in all circumstances. Making this point in Kantian terms, Cassirer claims that the moral judgments and theories that are more objective in nature are ones that are based not on a "'hypothetical rule'," that is, one that only holds under certain conditions, but on a "'categorical imperative'" that is unconditionally binding (AH 67).

Pursuing the topic of objectivity in the context of moral theory, Cassirer argues that we can trace a "'phenomenology of ethical consciousness'," which tracks the progress in moral philosophy toward theories that articulate increasingly universal principles (AH 79). To this end, he claims Greek philosophy advances toward Stoic ethics as a "prototype of ethical universalism" (AH 78). According to Cassirer, it is, in particular, the recognition of the "ideal of 'inner freedom'," as a normative principle that governs all human beings, that gives Stoic ethics its universality (AH 79). He then traces the development of this 'proto' ethical universalism into the full-blown ethical universalism that culminates in Kant's moral philosophy, which turns on "Kant's concept of duty and his concept of ethical autonomy" (AH 79). By Cassirer's lights, Kantian moral philosophy thus represents the most 'objective' moral theory because it offers a unified and universal account of the categorical imperative that governs the domain of willing. And it is these same universal principles of 'duty' and 'autonomy' that Cassirer thinks ground our more objective moral judgments.

However, in addition to offering this account of the cognition and objectivity involved in the 'factual conduct' of ethical consciousness, Cassirer provides an account of the conditions of its possibility. Disagreeing with Hägerström's empirical derivation of ethical consciousness on the basis of mechanical association, Cassirer argues that our ethical consciousness has its source in a function that is built into the *a priori* structure of the 'will', viz., the function tied to the ability to "actively bind-oneself [Sich-Binden]" and to make a 'promise' (Versprechen) (AH 104, 100). To actively bind oneself to a principle amounts to committing oneself to a principle in an autonomous, rather than heteronomous way. That is to say, this commitment involves self-determination, rather than determination by anything external. And, on Cassirer's view, this self-binding involves a 'promise' in the sense that we commit ourselves to choosing and acting on the basis of this principle in the 'future' (AH 100). According to Cassirer, the ability to autonomously bind oneself and to make promises provides ethical consciousness with a "fundamental direction" (Grundrichtung), which makes possible the 'factual conduct' of ethical consciousness (AH 104).

Thus, in his account of ethical consciousness, Cassirer defends an account of the 'factual conduct' of ethical consciousness that points toward the cognitivist nature of moral evaluation, the objectivity attainable in moral judgment and moral theory, and a phenomenology of ethical consciousness in moral philosophy. And he takes this all to be grounded in the *a priori* structure of the will and our ability to autonomously bind ourselves and make promises.

Right as a symbolic form

With an account of ethical consciousness in place, in *Axel Hägerström* Cassirer goes on to develop an account of right as a symbolic form. Guided by the transcendental method, Cassirer takes as his starting point "right as a cultural fact [*Kulturfaktum*]," that is, right as it is investigated in history, sociology, and anthropology (AH 102).[7] And he then endeavors to offer an account of the conditions that make this fact possible. Part of his analysis of these conditions turns on the account of how the *a priori* structures of the will just discussed, viz., our ability to autonomously bind ourselves and make promises, make the facts of right possible (see AH 100, 104). And the other part of his analysis of these conditions involves an account of right as a symbolic form. As a symbolic form, Cassirer treats right as "unique function of *objectification* [*eigentümliche Funktion der Objektivierung*]" through which we give a certain fixed and stable shape to our social relations and interactions (AH 102).

In Cassirer's account of the symbolic form of right, he treats it as a practical analogue of natural science. As I discussed in Chapter Six, he aligns both of these symbolic forms with the function of pure signification because they provide us with cognition of ideal concepts, laws, and principles. And he claims that whereas in natural science, we pursue theoretical cognition of nature, 'scientific experience', and the 'realm of being', in right we pursue practical cognition of the 'world of willing and action', 'social experience', and the 'realm of ought (*Sollen*)'.

Moreover, as is the case in natural science, Cassirer argues that progress in the symbolic form of right is measured in terms of our advancement towards theories and practices of right that are more 'unified' and 'universal' in character. As he puts it, in the symbolic form of right we advance from theories and practices that are applicable only to a "relatively narrow sphere [*Kreis*]" of social experience and the world of willing, for example, to a family or tribe, to ones that are applicable to "wider spheres [Kreisen]," for example, to a state and, eventually, to the world as a whole (AH 98).[8] Though Cassirer argues that Roman jurisprudence represents a significant step in this direction (AH 89–91, 98–99), he ultimately identifies a liberal approach to natural right (Naturrecht) as the most unified and universal approach to right (AH 100–102).

In Axel Hägerström Cassirer offers a sketch of a liberal account of natural right that he discusses in more detail elsewhere, for example, in Freedom and Form, "The Idea of the Republican Constitution," "On the Nature and Development of Natural Right," and Ch. 8 of The Myth of the State. As we shall see, this account turns not only on a particular understanding of natural right, but also an account of the social contract and idea of a 'republican constitution' that follows.

Cassirer traces, what he calls, the 'classical' account of natural right that he is sympathetic to back to thinkers, like Hugo Grotius, Samuel von Pufendorf, Leibniz, Christian Wolff, John Locke, William Blackstone, Montesquieu, and Voltaire (NDNR 218).[9] According to Cassirer, the "fundamental thesis [Grundthese]" of this tradition is that natural right has its 'source' not in anything 'outside' the human being, but rather something 'inside' the human being, viz., her 'autonomy' (Selbstgesetzlichkeit) and "spontaneity of spirit [Spontaneität des Geistes]" (NDNR 207).[10] So understood, natural right is something 'innate', which "springs directly from the nature [Wesen] of humanity, from its concept and essence," and is to be distinguished from 'acquired' (erworbene) rights, which have their source in "accidental determinations," like 'ancestry' or 'class' (IRC 9/297, translation modified). Unlike these 'acquired' rights that vary from person to person, since natural right has its source in the autonomy of the human being, this tradition treats it as a universal right, which is "valid and necessary . . . for all times and for all willing subjects [willenden Subjekte]" (NDNR 209). For this reason, specific natural rights, like the "right of equality" and the "right of personal security," that is, to "the unhindered exercise of all actions, upon which [a person's] survival as a physical being [physisches Wesen] and perfection as a spiritual being [geistige Wesen] depends," are understood as inalienable rights that every human being has (IRC 9/298, translation modified).

Cassirer, in turn, emphasizes the implications that this classical view of natural right has for how we understand the social contract. From the classical point of view, since natural rights are inalienable, they can be neither taken away, nor given away through any kind of agreement. This being the case, the classical view rejects any account of the social contract, according to which it involves individuals agreeing to give up their natural rights to, say, the sovereignty of the state (Hobbes) or the general will of the people (Rousseau) (NDNR 218). Given the inalienable status of natural rights in the classical tradition, Cassirer claims that they serve as a 'limit' on any social contract:

> The contract of rulership which is the legal basis of all civil power has, therefore, its inherent limits. There is no *pactum subjectionis*, no act of submission by which man can give up the state of a free agent and enslave himself. For by such an act of renunciation he would give up that very character which constitutes his nature and essence: he would lose his humanity.
>
> (MS 175/174)

According to Cassirer, the approach to the social contract that does the most justice to the classical concept of natural right is the social contract that Kant defends. As Cassirer emphasizes, on Kant's view, the social contract should not be understood in terms of any 'historical' agreement; it should be understood as an "original contract," which is ideal in nature (see AH 100–101, IRC 13/302–303, see Kant's "On the Common Saying" 8:297, "Perpetual Peace" 8:350, *Metaphysics of Morals* 6:315). This 'original contract' represents an ideal principle that governs the 'legitimacy' (*Rechtmäßigkeit*) of legislation (IRC 13/303). And this principle of legitimacy demands of legislators that they "give laws in such a way that could have arisen from the unified will of an entire people," qua autonomous human beings (see AH 100–101, IRC 13/302–302). For Kant, and Cassirer following him, this 'original' contract is politically realized through a "republican constitution," like the Declaration of Independence, which acknowledges that all human beings are "created by nature equally free and independent" and possess certain "inalienable rights" to life, liberty, and the pursuit of happiness, which cannot be taken away when they 'enter' into the state (IRC 10/298, translation modified). And this constitution, in turn, lays the groundwork for developing legitimate political and legal institutions that recognize and protect those natural rights.

By Cassirer's lights, insofar as this classical liberal approach to natural right, the social contract, and the republican constitution provides cognition of the ideal concepts, laws, and principles that govern the 'ought' of 'social experience' and the 'world of willing' in a unified and universal way, he treats it as the most 'objective' approach to right on offer. And in this achievement, Cassirer claims that the symbolic form of right contributes to our practical con- sciousness of freedom in unprecedented ways. On his view, the classical approach to right builds on an ethical consciousness that endorses unified and universal normative principles grounded in the recognition of the autonomy and equality of human beings. What is more, he regards this approach as one that treats that autonomy as the basis of our natural rights and as a limit on any social contract, and it enshrines and protects those rights via a repub- lican constitution and the legitimate political and legal institutions that are in keeping with that constitution. Thus, on Cassirer's view, even though other symbolic forms give us a sense of who we are as agents, and even, as morally responsible agents, none of them pro- vide us the awareness of this agency as a kind of autonomy, which is itself the source of the normative principles that we bind ourselves to in the domains of morality and right. More so than any other symbolic form, then, Cassirer argues that right involves a full prac- tical recognition, at both the individual and collective level, of the autonomy and spontaneity of human beings.

In these achievements, the symbolic form of right appears to be the practical counterpart of mathematics and natural science atop the 'ladder' of culture. Whereas mathematics and natural science advance our theoretical recognition of freedom, right advances our practical recognition of freedom to a higher degree than the other symbolic forms. However, Cassirer is clear that the achievements in both symbolic forms should be understood as achievements in different domains of culture. And he thinks that this is important to stress because otherwise it might seem as if there is a kind of competition between the two. For example, it might seem as if the concept of autonomy, championed by right, conflicts with the concept of causality operative in natural science. However, according to Cassirer, at the most advanced stages, both symbolic forms endorse a similar conception of causality, viz., as "orderliness according to law [Gesetzlichkeit]" (DI 206/248). And each symbolic form provides cognition of a different kind of 'orderliness according to law': right does so with respect to the world of willing and the 'realm of ought', whereas natural science does so with respect to nature and a 'realm of being' (DI 205/246). For this reason, he claims,

> [autonomy is] based on an independent *type* of orderliness according to law [*Typus von Gesetzlichkeit*], the autonomous orderly structure of the will. If this is kept in mind, it becomes understandable why ethics has nothing to fear and little to hope from the changes in the basic concepts of science which have taken place in modern physics.
>
> (DI 206/247, emphasis in original)

Hence his conclusion that our ethical and natural concepts of causality, "do not conflict because they belong to entirely different 'dimensions' of consideration. Thus they can never meet in one point; they cannot become identical, nor do they disturb or destroy each other" (DI 205/246). Thus, from Cassirer's perspective, rather than autonomy being in conflict with the conception of causality we have in natural science, they represent two variations on the same theme, viz., orderliness according to law.

For Cassirer, then, the achievements of right and natural science are not in competition with one another. Indeed, he treats them as complementary: whereas natural science, along with mathematics, provides us with the theoretical cognition needed to advance our theoretical consciousness of freedom, right gives us the practical cognition required to advance our practical consciousness of freedom. On Cassirer's view, it is thus mathematical, natural science, and right that sit on top of the ladder of culture, as the symbolic forms that teleologically advance us closest to the goal of consciousness of freedom.

The return of myth

In my discussion of Cassirer's philosophy of culture so far, I have concentrated primarily on his positive account of how culture teleologically progresses towards freedom. However, the political events of the 1930s and 1940s left little room for such optimism:

> What we have learned in the hard school of our modern political life is the fact that human culture is by no means the firmly established thing that we once supposed it to be . . . [W]e have to look upon the great master works of human culture in a much humbler way. They are not eternal nor unassailable. Our science, our poetry, our art, and our religion are only the upper layer of a much older stratum that reaches down to a great depth. We must always be prepared for violent concussions that may shake our cultural world and our social order to its very foundations.
>
> (MS 297/293)

To be sure, Cassirer never denies the possibility that culture can suffer 'violent concussions'. Indeed, he contrasts his view of the teleology of culture with Hegel's on precisely this point, rejecting what he takes to be Hegel's commitment to the necessity of the forward moving progress of culture, and endorsing an account of the contingency of this progress instead (see Chapter Five). Yet, positing the possibility of the regression of culture and witnessing its actualization are two different things. And the latter indelibly shaped the course not just of Cassirer's life as he went into exile, but of his philosophy of culture as well, as he directed his attention increasingly towards an analysis of the relationship between culture and the political crisis of the 1930s and 1940s.

Two questions, in particular, are of central interest to him in this period: how should we understand the cultural roots of the rise of fascism in the 20th century and what should be done in the face of it? In this section I focus on the first question and his diagnosis of the political situation in terms of the resurgence of myth, reserving a discussion of the proper response for the following section.

Cassirer opens *The Myth of the State* with the question, "How can we account for the new phenomenon [of myth] that so suddenly appeared on our political horizon and in a sense seemed to reverse all our former ideas of the character of our intellectual and social life?" (MS 3/7). And his core thesis is that the rise of fascism in the 1930s and 1940s is the result of "the preponderance of mythical thought over rational thought" (MS 3/7). Analyzing this thesis will not only shed light on Cassirer's view of the fragility of cultural progress, but also his account of the particular conditions that allowed for myth to again gain ascendency.

Recall that for all his insistence on cultural progress, Cassirer balances the 'ladder' metaphor for culture with the 'bow-and-lyre' metaphor, which reflects the distinctive contribution that each symbolic form makes to culture (Chapter Five). And, on Cassirer's view, both metaphors reflect an on-going truth about culture: though culture can progress, this does not mean that less teleologically advanced symbolic forms are 'sublated' into more teleologically advanced symbolic forms. Rather, even at a more advanced stage in culture, Cassirer claims that the less teleologically advanced symbolic forms remain an autonomous force within culture.

What this means for Cassirer's analysis of myth is that even when culture takes steps beyond the dominance of myth, this progress never involves the elimination of myth altogether; myth, instead, continues to exist as a form through which we symbolically shape our cultural world. And in many ways, Cassirer thinks that this is a good thing:

> [Myth] still has its place in the most advanced stages of human culture. To banish myth, to eradicate it root and branch, would mean an impoverishment. What would become of poetry and art if we would no longer be able to understand the language of myth. . . . Then we would have no access to the world of our great poets, painters, and sculptors—to the works of Aeschylus, Pindar, or Pheidias, to the works of Dante, Shakespeare, Milton, or Michelangelo.
>
> (TMPM 245)

However, Cassirer claims that in order for myth to play a positive role in a more teleologically advanced stage of culture, it must "be counterbalanced and brought under control" by the other symbolic forms (TMPM 246). That is to say, myth should not be the dominant form in accordance with which we understand who we are and the world around us; instead, we should balance the mythical perspective with the perspective from the other forms in a way that is in keeping with our consciousness of freedom.

Nevertheless, Cassirer indicates that this equilibrium is never guaranteed:

> myth has not been really vanquished and subjugated. It is always there, lurking in the dark and waiting for its hour and opportunity. This hour comes as soon as the other binding forces of man's social life, for one reason or another, lose their strength and are no longer able to combat the demonic mythical powers.
>
> (MS 280/275)

On Cassirer's view, if conditions arise in which the other forces of culture do not counterbalance myth, then it becomes possible for mythical thinking to become the dominant force by means of which we understand and organize ourselves and the world.

According to Cassirer, with the rise of fascism in the 1930s and 1940s, this possibility became reality, as mythical thought came to dominate rational thought. And in his political writings, he endeavors to offer an account of the conditions that made this resurgence of mythical thinking possible. He points towards two factors, one more direct, viz., modern political techniques, and one less direct, viz., the complicity of philosophers.

Cassirer describes the fascist political program that most directly precipitated the return to a mythical outlook as the 'technique of myth'. As he sees it, fascist governments developed a program to reintroduce certain myths through technological means and this helped inculcate a more mythical mode of thinking among its citizens.

On Cassirer's assessment, there were three myths in particular that were revived through this technical program: the myth of hero worship, race worship, and state worship. Though these myths all have a long history, Cassirer suggests that their reformulation in the 19th century paved the way for their appropriation within the fascist framework. Beginning with the myth of the hero, Cassirer identifies the relevant source as Carlyle's lectures "On Heroes, Hero Worship and the Heroic in History" (1840), in which he defends the idea that there are exceptional men, heroes, who should serve as the leaders of ordinary men (see MS Ch. 15). While Cassirer insists that Carlyle himself defines the hero in moral terms and would therefore not have condoned fascist ideology, he argues that Carlyle's view nevertheless prepared the way for the fascist view of the dictator. For stripped of any moral obligation, the hero emerges as the supreme leader, who alone has authority: "The former social bonds—law, justice, constitutions—are declared to be without any value. What alone remains is the mystical power and authority of the leader and the leader's will is supreme law" (MS 280/276).

Meanwhile, with respect to the race worship, Cassirer argues that it is Gobineau's *Essay on the Inequality of the Human Races* (1853–1855), which plays a central role (see MS Ch. 16). On Cassirer's interpretation, what is potent in the *Essay* is his "theory of the 'totalitarian race'" (MS 231/228). According to Cassirer's reading, Gobineau's theory of the totalitarian race involves a white supremacist commitment to there being only one independent value that all other values are derived from, viz., the white race. According to Cassirer, this theory "marked the road" to the ideology of the fascist state, as the latter took itself to be justified as the representative of the supreme white race (MS 232/228).

The third myth, the myth of state worship, is one that Cassirer traces back to Hegel (see MS Ch. 17). Though as we shall see shortly, Cassirer does not think that Hegel himself would endorse the fascist governments of the 20th century, nevertheless he asserts that, "No other philosophical system has done so much for the preparation of fascism and imperialism as Hegel's doctrine of the state" (MS 273/270). By Cassirer's lights, it is in particular Hegel's state-absolutism, that is, his emphasis on the supreme value and power of the state, that was appropriated to fascist ends. As Cassirer points out, Hegel rejects the Kantian picture of morality and politics as most fundamentally grounded in the autonomy and inalienable rights of individuals as an overly 'subjective' conception, favoring instead a view of the state as the proper 'objective' source (MS 263/260). Cassirer takes this latter view to be evident in Hegel's claims that the state is "the supreme and most perfect reality" because it is "the very incarnation of the 'spirit of the world,'" and "the 'Divine Idea as it exists on earth'" (MS 263/260). Furthermore, Cassirer points out that Hegel asserts, "'Men are as foolish as to forget . . . in their enthusiasm for liberty of conscience and political freedom, the truth which lies in power'" (MS 267/264 quoting Hegel). According to Cassirer, it is this last statement in particular that "contain[s] the clearest and most ruthless program of fascism that has ever been propounded by any political or philosophic writer" because these words attribute value and truth not just to the state, but to the *power* of the state (MS 267/264).

This being said, as was the case with Carlyle, Cassirer takes care to point out that Hegel himself would not have supported the modern fascist state. To begin, Cassirer claims that Hegel recognizes that there are "forms of cultural life [that] have an independent meaning and value" outside the state, viz., art, religion, and philosophy, and that the "highest aim" of the state should be to promote these cultural forms (MS 275/271). Moreover, Cassirer argues that whereas fascist ideology adheres to "the principle of *Gleichschaltung* [enforced conformity]," which levels out all social and cultural differences, Hegel recognizes that freedom can only be achieved in difference, in the dialectical encounter of opposites (MS 275/272). Though Cassirer thinks it is crucial to recognize these differences between Hegel's view and the fascist program, he nevertheless argues that there are themes in Hegel's view of state-absolutism, which provided the basis for a new fascist version of the myth of the state.

According to Cassirer, although these three myths of the hero, race, and state were familiar in the 19th century, in the 20th century fascist governments developed a technique for these myths that turned them into weapons:

> To change the old ideas into strong and powerful political weapons something more was needed. . . . A new technique had to be developed. This was the last and decisive factor. To put it into scientific terminology we may say that this technique had a catalytic effect. It accelerated all reactions and gave them their full effect. While the soil for the Myth of the Twentieth Century

had been prepared long before, it could not have borne its fruit without the skillful use of the new technical tool.

(MS 277/273)

Indeed, he compares the manufacturing of myth to the manufacturing of machine guns and airplanes, claiming that they are simply two kinds of weapons, intellectual and material, produced "for the same purpose, for internal and external warfare" (TMPM 253, see also MS 282/277–278).

In order to illustrate how fascist governments developed techniques to inculcate this mythical thinking, Cassirer points to examples like the language of Nazi propaganda and the institution of new rites. With regard to propaganda, Cassirer argues that the Nazis relied on the use of 'magic' words in their mass-marketed propaganda, that is, words the purpose of which was "to produce certain effects and to stir up certain emotions" (MS 283/278). By both coining new terms and transforming old ones, Cassirer claims that the words used in Nazi propaganda became a tool for promoting their ideology, viz., by stirring up negative emotions, like hatred, fear, and anger towards people who did not belong to the Aryan race, and positive emotions, like those connected to supremacy, strength, and triumph, for those belonging to the Aryan race.

The second example Cassirer mentions involves the employment of new rites. As Cassirer points out, rites become a pervasive aspect of the fascist program, as all actions came to have a ritual element about them and special rites were developed for members of different classes, genders, religions, and age groups (see, e.g., MS 284/279–280). By connecting these new rituals to concepts familiar from older mythical rituals, like the concept of the scapegoat or a blood-ritual, Cassirer claims that they became affectively charged weapons for the oppression of allegedly inferior races. From his perspective, it is the development of these sorts of techniques, among others, that made the reintroduction of the myth of the hero, race, and state so effective and that contributed to the regression into a mythical mode of thought.

According to Cassirer, the ultimate effect of the technique of myth was to inculcate a more mythical mindset, dominated by collective thought (see PP 231–232, AS 322–323). In this vein, he claims that the technique of myth turns on a reconceptualization of the moral subject in collective terms: "If there is any 'moral' subject—the community, the nation, the race are held answerable for its actions. The acts are good or evil according as they are done by a superrace or by an inferior race" ("Judaism and Modern Political Myths" 203). Cassirer maintains that this collective mentality promotes a willingness of individuals to let their beliefs and choices be determined collectively, "by nationality, creed, political party, social position, and other elements in one's surroundings" (AS 322). And in the fascist context, this collective mentality was oriented by the myths of hero worship, white supremacy, and the absolute power of the state.

On Cassirer's assessment, what this mythical mode of thinking fundamentally encourages is the relinquishing of any sense of autonomy and individual responsibility in favor of a more passive conception of the self as determined by forces outside her control. As Cassirer describes this transformation,

> We have learned that modern man . . . can easily be thrown back to a state of complete acquiescence. . . . Of all the sad experiences of these last twelve years [from 1933–1945] this is perhaps the most dreadful one. It may be compared to the experience of Odysseus on the island of Circe. But it is even worse. Circe had transformed the friends and companions of Odysseus into various animal shapes. But here are men, men of education and intelligence, honest and upright men who suddenly give up the highest human privilege. They have ceased to be free and personal agents. . . . In fact they are moved by an external force. They act like marionettes in a puppet show.
>
> (MS 286/281)

By Cassirer's lights, this regression in the consciousness of freedom is a price of the return of mythical thinking in the 20th century.

Although Cassirer thinks that the technique of myth is directly responsible for the resurgence in mythical thinking, given his home discipline, he also considers the extent to which some philosophers might have indirectly contributed to this resurgence. To this end, he discusses particular philosophers that he thinks are in some way culpable for promoting a mythical mode of thinking, as well as the duties that philosophers more generally failed to live up to.

Beginning with his more specific criticisms, Cassirer singles out Oswald Spengler and Heidegger as two philosophers who he thinks contributed, albeit it in indirect ways, to the cultivation of a mythical mentality (see MS 289–293/284–288, PP 227–230). In *The Decline of the West* (1918), Spengler argues that the decline of Western civilization is at hand and that there is nothing any single individual can do about it because it is the West's 'destiny' and 'fate'. On Cassirer's analysis, in promoting the concepts of 'destiny' and 'fate', Spengler revives ancient mythical concepts, which encourage a passive conception of the self as subject to forces outside her control.

Meanwhile, as I emphasized in Chapter One, Cassirer's long-standing criticism of Heidegger focuses on Heidegger's conception of the human being as defined in terms of her *Geworfenheit*, her 'thrownness', rather than in terms of her spontaneity. On Cassirer's gloss, by *Geworfenheit* Heidegger has in mind the idea that individuals are "thrown into the stream of time," that is, into a particular set of historical and temporal conditions, and that this stream is one that we are at the mercy of: "We cannot emerge from this stream and we cannot change its course. We have to accept the historical conditions of our existence. We can try and understand and to interpret them; but we cannot change them" (MS 293/287–288). In defining the human being in terms of her thrownness, Cassirer charges Heidegger, like Spengler, with the mistake of promoting a passive conception of the human being.

In assessing the influence of Spengler and Heidegger, Cassirer denies that either have "a direct bearing on the development of political ideas in Germany" because these ideas arise from "different sources" (MS 293/288). Nevertheless, he asserts that their philosophies "did enfeeble and slowly undermine the forces that could have resisted the modern political myths" (MS 293/288). By his lights, they did so because they failed to uphold their 'educational' duty as philosophers:

> As soon as philosophy . . . gives way to a merely passive attitude, it can no longer fulfill its most important educational task. It cannot teach man how to develop his active faculties in order to

form his individual and social life. A philosophy that indulges in somber predictions about the decline and the inevitable destruction of human culture, a philosophy whose whole attention is focused on the *Geworfenheit*, the Being-thrown of man, can no longer do its duty.

(PP 230)

As philosophers, Cassirer claims that Spengler and Heidegger failed in their educational obligation to promote an active and autonomous understanding of the human being, an understanding Cassirer sees as a 'force that could have resisted the modern political myths'. And it is in this sense that Cassirer thinks Spengler and Heidegger contribute in a certain way to the resurgence of mythical thinking.

However, in addition to this criticism of Spengler and Heidegger, Cassirer considers other ways in which philosophers may have fallen short in the 1930s and 1940s. And he argues that many philosophers, himself included, had an obligation to protect culture, which they did not fulfill. Quoting Schweitzer to this end, Cassirer says,

"Every effort should have been made to direct the attention of the educated and the uneducated to the problem of cultural ideals. . . . But in the hour of peril the watchman slept, who should have kept watch over us. So it happened that we did not struggle for our culture."

(CP 60/155–6, see also PP 232)

Continuing then with his own words, Cassirer states,

I believe that all of us who have worked in the area of theoretical philosophy in the last decades deserve in a certain sense this reproach of Schweitzer; I do not exclude myself and I do not absolve myself. While endeavoring on behalf of the scholastic conception of philosophy, immersed in its difficulties as it caught in its subtle problems, we have all too frequently lost sight of the true connection of philosophy with the world.

(CP 60/156)

In this passage, Cassirer alludes to two conceptions of philosophy: a 'scholastic' conception, according to which the aim of philosophy is to solve abstract, theoretical problems, and a 'cosmopolitan' conception, according to which the true task of philosophy is a practically oriented one, that requires philosophy be "related to the world" and our concrete place in it (CP 58–59/153–154). From Cassirer's perspective, in a more scholastic rather than a cosmopolitan direction, many philosophers, himself among them, neglected their duties as the 'watchmen of culture' and so failed to 'struggle for culture', at a moment when it was needed.

The ethical struggle for culture

So what does Cassirer think should be done to combat the resurgence of myth? To be sure, part of Cassirer's answer involves the need to return to the liberal political and juridical structures, which protect and enshrine the autonomy, equality, and inalienable rights of all human beings, which I discussed above. However, what I shall focus on below is the part of Cassirer's answer that turns on an account of what philosophers and human beings should do on a more personal level.

In assessing the role of philosophy in this context, as we just saw, Cassirer claims that philosophers have certain 'duties'. These duties include the educational duty to "teach man how to develop his active faculties in order to form his individual and social life," that is, to help people understand not only that they are autonomous, but also how to develop this autonomy at an individual and cultural level (PP 220). And they also include the cosmopolitan duty of philosophy to engage in thought that relates to the world and so bears on the actual, concrete situations that we find ourselves in.

In addition to this educational and cosmopolitan duty, however, Cassirer argues that philosophers have a further duty to "understand the adversary" (MS 296/290). Indeed, Cassirer thinks that there was a widespread failure to take the resurgence of mythical thinking seriously:

> When we first heard of the political myths we found them so absurd and incongruous, so fantastic and ludicrous that we could hardly be prevailed upon to take them seriously. By now it has become clear to us all that this was a great mistake.
>
> (MS 296/291)

According to Cassirer, philosophers can help contribute to this oversight by attempting to 'understand the adversary' by devoting philosophical attention to "the origin, the structure, the methods, and the technique" of myth (MS 296/291).

In describing these three philosophical duties as 'duties', Cassirer means to emphasize their moral status. On his view, they reflect the moral obligations that philosophers have to recognize and support the autonomy that individuals have and that culture is an expression of. So understood, Cassirer suggests that these duties are ones that philosophers ought to fulfill regardless of whether they want to or not. Indeed, he acknowledges that for some philosophers, especially those interested in theoretical and abstract issues, engaging in this kind of philosophy is "no very attractive task"; however, he insists, in a Kantian vein, that it must nevertheless be "done from duty" (TMPM 266–267). By Cassirer's lights, then, philosophers have a moral duty to contribute to the understanding and realization of our autonomy, as it is expressed on an individual and collective level, and the forces and powers that threaten to undermine it. This moral role, he claims, is ultimately "the function which philosophy has to perform in the whole of our social life" (TMPM 267).

However, in addition to the moral obligations that Cassirer details for philosophers, he also argues that we as human beings have moral duties. At the most basic level, Cassirer argues that this amounts to the responsibility to recognize and act on our autonomy. However, according to Cassirer, we need to grasp our autonomy not as something that is simply 'given' (*gegeben*); it is '*aufgegeben*', given as a task that we must constantly strive towards:

> freedom . . . is not a fact but a postulate. It is not *gegeben* but *aufgegeben*; it is not a gift with which human nature is endowed; it is rather a task, and the most arduous task man can set himself. It is . . . an ethical imperative.
>
> (MS 287–288/282).

For Cassirer, this conception of freedom as an ethical task has implications at both the individual and cultural level. At the individual level, he suggests that each of us is responsible for realizing this freedom in our own lives, by striving to form our beliefs, make our decisions, and act in autonomous ways, rather than on a heteron-omous basis, like collective thought. Cassirer acknowledges that this is a burden: it is "easier to depend upon others than to think, to judge, and to decide for himself," all the more so, he suggests, in difficult social, political, and economic situations (MS 288/283). Nevertheless, he claims that we still have the ethical obligation to take up the burden of our freedom and take responsibility for our own lives.

Yet in addition to this individual responsibility, Cassirer maintains that our efforts towards freedom must manifest on the cultural level. Recall that Cassirer conceives of the "problem of freedom" as the "ethical problem of culture" (CIPC 82/112). Thus, on his view, we have an ethical responsibility to build and preserve the cultural world, not just as the expression of our freedom, but as the path towards our consciousness of freedom. In his words,

> [Culture] is not a given thing; it is an idea and an ideal. It must be understood in a dynamic sense, instead of conceiving it in a static sense. It must be produced, and in this production consists the essential meaning of culture and its ethical value.
>
> (CIPC 90/118)

On Cassirer's view, then, we have an ethical responsibility to actively contribute to and struggle for culture, as the outward realization of our freedom.

Summary

In this chapter, I have explored the ethical and political dimensions of Cassirer's philosophy of culture and concluded my analysis of his system of individual symbolic forms. I opened with a discussion of Cassirer's explicit extension of The Philosophy of Symbolic Forms in practical directions in Axel Hägerström. There, I considered Cassirer's rejection of a non-cognitivist, relativist view of ethical conscious-ness in favor of a cognitivist account, which treats moral evaluations or opinions as judgments and which defines the standard of moral objectivity in terms of the 'unity' and 'universality' of our normative principles. I also touched on Cassirer's account of the 'phenomen-ology of ethical consciousness', which he glosses in terms of the development of moral theories toward the ideal of objectivity and the categorical imperatives tied to the autonomy of human beings. In addition to clarifying these 'factual' features of ethical consciousness, I examined Cassirer's claim that this is all made possible by the *a priori* structures of the will, viz., our ability to autonomously bind ourselves and make promises.

I next addressed Cassirer's account of right as a symbolic form that he aligns with the function of pure signification and which provides practical cognition of the ideal concepts, laws, and principles that govern the oughts of the 'world of willing' and 'social experience'. I highlighted Cassirer's commitment to the 'classical' liberal conception of natural right, the social contract, and the republican constitution. And I explored the ways in which Cassirer thinks that this advances our practical recognition of freedom, enabling us not only to be fully conscious of the autonomy of human beings, but also to collectively secure and protect this autonomy through legitimate political and juridical means. By way of concluding this discussion of right, I analyzed the teleological parallel Cassirer sees between right, on the one hand, and mathematics and natural science, on the other. To this end, I examined Cassirer's view that these symbolic forms advance culture closest to the goal of culture, viz., consciousness of freedom—right doing so through our practical recognition of freedom, and mathematics and natural science doing so through a theoretical recognition of freedom. And it is for these reasons that I claimed that, for Cassirer, right belongs atop the ladder of culture alongside mathematics and natural science.

However, although Cassirer thus tracks the progress of culture along these lines, I also considered his account of the regression of culture, specifically as it manifested in the rise of fascism in the 20th century. I discussed Cassirer's general account of the fragility of culture and his diagnosis of the rise of fascism in the 1930s and 1940s as the direct result of the 'technique of myth' deployed by fascist governments and as the indirect result of philosophers falling short of their duties. I also addressed Cassirer's account of what, in addition to a return to classical liberal political and juridical practices, should be done in light of this regression. There, I examined Cassirer's account of specific moral duties philosophers have to educate, to make their philosophy relevant to the world, and to understand the adversary. I also examined his account of the moral obligations that we all have to not only acknowledge our autonomy at an individual level, but also to struggle for the realization of our autonomy in our cultural world.

Thus, far from neglecting moral and political philosophy, in the 1930s and 1940s Cassirer expands his philosophy of culture in order to account for the nature of ethical consciousness, the symbolic form of right, and the rise of fascism. Indeed, by developing his thought in these directions, Cassirer with all his Olympian tendencies, strives to live up to the ethical duties he thinks we all have to safeguard culture.

Notes

1 Strauss continues by saying that a moral transformation of the philosophy of symbolic forms would involve "a return to Cassirer's teacher Hermann Cohen, if not to Kant himself" (Strauss 1947: 128). Making a similar point elsewhere, Strauss claims, "Having been a disciple of Hermann Cohen he had transformed Cohen's philosophic system, the very center of which was ethics, into a philosophy of symbolic forms in which ethics has silently disappeared" (Strauss 1988: 246). For a discussion of Strauss's relationship to Cassirer (and Heidegger), see Gordon (2010): 315–319, 343–345.

2 Cassirer does not appear to observe a strict distinction between 'ethics' (Sittlichkeit) and 'morality' (Moralität), so I will use them interchangeably below.

3 Cassirer delivered this at a celebration of the 10th anniversary of the Weimar Republic.

4 Cassirer lists morality and right as symbolic forms at PSFv2 xxx/xi/xiv, LM 168/266 (which he cites again in AH 82 fn. 1), FT 275/142. See Freudenthal (2008) for the argument that morality is not a symbolic form.

5 See also his claim that he wants to use Chapter Four to offer a 'sketch' that will fill in the 'gap' in the philosophy of symbolic forms with regard to the realm of right (AH 82, fn. 1).

6 This is a modification of Cato's "ceterum censeo Carthaginem esse delendam" ("furthermore I propose that Carthage must be destroyed").

7 Cassirer pays especially close attention to the facts of Roman jurisprudence because this is a topic that Hägerström devotes a great deal of attention to.

8 For his discussion of progress in natural science in similar terms, see AH 97–98, SF 271–281.

9 Translations of NDNR are my own.

10 In this context, Cassirer compares the 'ideal' origin of the concept of natural right to the origin of mathematical concepts (NDNR 204–206).

Further reading

Bottici, Chiara (2007). *A Philosophy of Political Myth*. New York: Cambridge University Press. See Chapter 8 for a critical discussion of Cassirer's philosophy of political myth and its relation to other figures, like Sorel and Spinoza.

Capeillères, Fabien (2015). "Cassirer on the 'Objectivity' of Evil: The Symbolic Constitution of *Der Mythus des 20 Jahrhunderts*." In *The Philosophy of Ernst Cassirer: A Novel Assessment*, edited by J. Tyler Friedman and Sebastian Luft, 435–468. Berlin: De Gruyter. A discussion of Cassirer's analysis of Nazism.

Cornell, Drucilla and Panfilio, Kenneth (2010). *Symbolic Forms for a New Humanity: Cultural and Racial Reconfigurations for a New Humanity*. New York: Fordham University Press. A discussion of how Cassirer's and Fanon's theories of symbolism can be integrated in a theory of social change and revolution. See further discussion in Chapter Eight.

Coskun, Deniz (2007). *Law as Symbolic Form: Ernst Cassirer and the Anthropocentric View of Law*. Law and Philosophy Library 82. Dordrecht: Springer. A discussion of Cassirer's ethical, juridical, and political commitments and a discussion of law as a symbolic form.

Curthoys, Ned (2013). *The Legacy of Liberal Judaism: Ernst Cassirer and Hannah Arendt's Hidden Conversation*. New York: Berghahn Books. A comparison of Cassirer's and Arendt's ethics in the framework of liberal Judaism.

Gordon, Peter (2010). *Continental Divide*. Cambridge: Harvard University Press. A discussion of the political context, content, and ramifications of the debate between Cassirer and Heidegger, highlighting, in particular, the difference their respective endorsements of the human being as spontaneous and thrown.

——(2012). "German Idealism and German Liberalism in the 1920s: The Case of Ernst Cassirer." In *The Weimar Moment: Liberalism, Political Theology, and Law*, edited by Leonard Kaplan and Rudy Koshar, 337–344. Lanham: Lexington Books. A discussion of "The Idea of a Republican Constitution" in the context of the affiliation of German Idealism with liberalism in the 1920s.

Krois, John Michael (1987). *Cassirer: Symbolic Forms and History*. New Haven: Yale University Press. See Chapter 4, "Morality and Law," for a discussion of Cassirer's account of morality and law.

Luft, Sebastian (2015). *The Space of Culture: Towards a Neo-Kantian Philosophy of Culture*. Oxford: Oxford University Press. See pp. 221–231 for a discussion of many of the key questions in the debate surrounding Cassirer's ethical commitments.

Mali, Joseph (2008). "The Myth of the State Revisited: Ernst Cassirer and Modern Political Theory." In *The Symbolic Construction of Reality: The Legacy of Ernst Cassirer*, edited by Jeffrey Andrew Barash, 135–162. Chicago: University of Chicago Press. A discussion of Cassirer's legacy in modern political theory with regard to the concept of myth.

Moynahan, Gregory (2013). *Ernst Cassirer and the Critical Science of Germany, 1899–1919*. London: Anthem Press. See Part III for a discussion of the trajectory of Cassirer's early thought as culminating in the model of political liberalism he develops in *Freedom and Form*.

Renz, Ursula (2011). "From Philosophy to Criticism of Myth: Cassirer's Concept of Myth." *Synthese* 179: 135–152. A discussion of the continuous development of Cassirer's analysis of myth from *The Philosophy of Symbolic Forms* to *The Myth of the State*.

——(2020). "Cassirer's Enlightenment: On Philosophy and the 'Denkform' of Reason." *British Journal for the History of Philosophy* 28(3): 636–652. A discussion of the way that Cassirer's political commitments manifest in his historical writings, specifically in *The Philosophy of Enlightenment*.

Rudolph, Enno, ed. (1999). *Cassirers Weg zur Philosophie der Politik*, Cassirer-Forschungen 5. Hamburg: Meiner. A volume of collected essays devoted to Cassirer's political philosophy.

Skidelsky, Edward (2008). *Ernst Cassirer: The Last Philosopher of Culture*. Princeton: Princeton University Press. See Chapter Nine, "Politics," for a critical discussion of Cassirer's relationship to the tradition of German liberalism and his political views on myth.

Verene, Donald (2011). *The Origins of the Philosophy of Symbolic Forms*. Evanston: Northwestern University Press. See Chapter Seven, "Human Freedom and Politics," for a discussion of politics as a symbolic form and the political dimensions of Cassirer's account of freedom and myth.

Eight
Cassirer's legacy

Introduction

Given the breadth of Cassirer's *oeuvre*, it should come as no surprise that his legacy extends in many directions, from philosophy to psychology, anthropology to art history, history to hermeneutics. In this chapter I aim to trace some, though by no means all, of Cassirer's lines of influence in the 20th and 21st centuries.

However, before doing so, a few words are in order regarding the complications that surround Cassirer's reception. Although during his lifetime, Cassirer was among the most prominent German philosophers, well regarded in Europe and the United States, after his death in 1945 interest in his work waned. There are various reasons for this. In the wake of World War II, Cassirer's defense of the Enlightenment ideals of humanity, freedom, and progress seemed outdated. Moreover, his thought was not easily situated in relation to the rift that developed in philosophy between the so-called 'analytic' tradition stemming from logical positivism and the so-called 'continental' tradition with its roots in existential-phenomenology. To be sure, Cassirer's work shared affinities with both traditions. Like the logical positivists, Cassirer was interested in offering an account of the logical and cognitive dimensions of mathematics and natural science. However, his idealist commitments to the Copernican Revolution conflicted with the empiricist orientation of these positivist thinkers. And in keeping with the existential-phenomenological focus of thinkers like Heidegger, Cassirer endeavored to provide an account of our everyday experience of the world, our selves, and others. Yet, as emerged in the Davos disputation with Heidegger, Cassirer's conception of the human being as fundamentally autonomous appeared to be at odds with the existential definition of the human being as thrown into the world. Thus, even though Cassirer's system made contact with both sides of the analytic-continental divide, it did not fall neatly on one side or the other, and was overshadowed accordingly.

Yet in spite of these complications in his reception, Cassirer's ideas have shaped generations of thinkers. During his lifetime and immediately after his death, his philosophical impact is evident in the work not only of his students, like Kurt Lewin and Arthur Pap, but also his colleagues, like Erwin Panofsky and Susanne Langer. Later in the 20th century, his ideas influenced a wide range of thinkers, including Peter Gay, Nelson Goodman, Maurice Merleau-Ponty, Hans Blumenberg, among others. Meanwhile, in the 21st century, his legacy continues to be carried forward, notably in the philosophy of science and critical theory.

In what follows, I focus on four aspects of his legacy. I start with a discussion of the influence of Cassirer on more recent debates in philosophy of science about structural realism and the dynamic *a priori*. In the following section I concentrate on the impact of Cassirer's ideas on psychology and social science. Next, I examine the importance of Cassirer's historical work for the field of intellectual history. In the remainder of the chapter I consider the lasting significance of Cassirer's philosophy of symbolic forms. To this end, I begin with the influence of his philosophy of symbolism in general on the philosophers, Langer, Goodman, and Merleau-Ponty. I then address its relevance for Panofsky's method for art history, Blumenberg's account of myth, the account of language developed by the Bakhtin Circle, and the critical theory of Jürgen Habermas, Drucilla Cornell, and Kenneth Michael Panfilio.

The legacy of Cassirer's philosophy of science

There has been a recent resurgence of interest in Cassirer's philosophy of science particularly with respect to contemporary debates about structural realism and the dynamic *a priori*.

Structural realism

One of the central debates in contemporary philosophy of science is whether we should be realists or antirealists about science. According to a scientific realist, we should be committed to the existence of the unobservable entities articulated by the most successful scientific theories. Often cited in support of realism is the 'no miracles' argument, according to which success in science is not a miracle and theories are, at least approximately, true descriptions of the world. Meanwhile, an antirealist remains skeptical about the existence of the unobservable entities described by scientific theories. An argument that is often mustered in favor of this position is the 'pessimistic meta-induction' argument, according to which, given that scientific theories of the past turned out to be false, by induction, we should conclude that our current best scientific theories will turn out to be false as well.

Structural realism is a view that attempts to cut a middle path between scientific realism and antirealism.[1] In keeping with traditional realism, a structural realist wants to preserve the idea that scientific theories are at least approximately true descriptions of the world. However, like the antirealist, a structural realist endorses the idea that we should be skeptical when it comes to making existence claims about what is unobservable. In an effort to do justice to both ideas, a structural realist argues that our current best scientific theories are approximately true accounts not of unobservable objects, but rather of structures.

Structural realism comes in many varieties, two of which are relevant here: epistemic structural realism (ESR) and ontic structural realism (OSR). According to ESR, we can only know the mathematical structures or relations between unobservable objects and so must refrain from making claims about the ultimate nature of those objects. In Kantian terms, we can only know the structure or relations between 'appearances' or 'phenomena', not the things in themselves. Unlike ESR, which is a theory about what we can know, OSR is an ontological theory about the nature of reality. According to OSR, at the most fundamental level, reality is constituted not by objects, but by structures. Hence the slogan that is associated with OSR: structure is all there is.

Proponents of both ESR and OSR have regarded Cassirer's philosophy of science as an early anticipation of structural realism. In the case of ESR, for example, Barry Gower, Michela Massimi, and Jeremy Heis have argued that the seeds of this view can be traced back to Cassirer's functional approach to science more generally and to specific claims like, "we never know things as they are for themselves, but only in their mutual relations," or "The determinations expressed by scientific concepts are not perceptible *properties* of the empirical objects . . . they are *relations* of these empirical objects" (SF 305/330, 229/250, emphasis in original).[2]

Meanwhile, defenders of OSR, like Steven French and James Ladyman, have suggested that aspects of Cassirer's view, especially his analysis of quantum mechanics in *Determinism and Indeterminism*, anticipates their own position.[3] These defenders take Cassirer's view to be amenable to the following OSR commitments:

a) Relations are conceptually prior to objects
b) The locus of objectivity shifts from objects to laws and symmetries (French (2014): 99).

Notice that no mention is here made of Cassirer's critical idealism and this is no mistake, for the proponents of OSR make clear that theirs is a modification of Cassirer's position insofar as they conceive of (a) and (b) as 'detached' from the Neo-Kantian idealism in which Cassirer presents them, and as reflecting, instead, the nature of a mind-independent reality (Cei and French (2009): 114, French (2014): 99).

This being said, reading Cassirer along the lines of both ESR and OSR has met with some resistance. For example, Heis and Thomas Mormann worry that interpreting Cassirer as a proponent of OSR covers over the critical idealism that motivates Cassirer's functional approach to science.[4] In a slightly different vein, Ryckman worries about OSR and, more generally, about any theory that would ascribe to Cassirer a metaphysical view insofar as he takes Cassirer's claims in *Determinism and Indeterminism* to the effect that the law is prior to the object to reveal that Cassirer's philosophy of science should, ultimately, be read "solely as an epistemology, not a metaphysics, structuralist or otherwise" (2015: 100, see DI

131/159). Meanwhile, raising objections to the move that would attribute to Cassirer either OSR or ESR, Heis argues that insofar as Cassirer rejects the traditional realist claim that scientific theories are true descriptions of the world, he lacks a key commitment that motivates structural realism (2014b: 18). This, in turn, leaves open the question of whether Cassirer would have endorsed any contemporary version of structural realism, epistemological, ontic, or otherwise.[5]

Cassirer on the dynamic a priori

In contemporary philosophy of science, Cassirer's views have also been influential with respect to discussions about the dynamic *a priori*. As I noted in Chapter Four, Cassirer rejects Kant's account of the *a priori* as something that is wholly fixed, in favor of an account of the *a priori* as something that can dynamically change over time. And this aspect of his account of the *a priori* has garnered attention, if little consensus from different quarters in the philosophy of science. To trace this line of influence, I shall first consider the immediate reception of Cassirer's theory of the *a priori* by Arthur Pap before turning to recent debates about how to interpret Cassirer's view.

During Cassirer's lifetime, Arthur Pap deployed Cassirer's account in his pragmatic analysis of what he calls the 'functional *a priori*'.[6] Arthur Pap was a Swiss Jew who, like Cassirer, had fled the Nazis and immigrated to the U.S., where he earned his Master's degree under Cassirer's supervision at Yale and then his doctorate at Columbia under Ernest Nagel. Pap's early efforts in the philosophy of science included a series of articles, "On the Meaning of Necessity" (1943a), "On the Meaning of Universality" (1943b), and "The Different Kinds of A Priori" (1944), and these culminated in The A Priori in Physical Theory (1946). In these writings, Pap takes aim at the logical empiricist denial of synthetic *a priori* knowledge in science, arguing that over and above the formal or analytic, *a priori* and the material or 'self-evident', *a priori*, there is a functional *a priori* that reflects not what is logically necessary, but rather what is 'hypothetically' or 'functionally' necessary in light of the ends or aims of a particular theory. As such, Pap claims these *a priori* statements are 'made' or 'selected' as necessary in light of the functional aims of a particular theory (1944: 479, 1943a: 451). And Pap treats this kind of *a priori* as dynamic in the sense that it will change as our theories do.

Although this way of construing the functional *a priori* has conventionalist and pragmatist overtones, which Pap traces back to Poincaré, Dewey, and C.I. Lewis, he also attributes this view of the *a priori* to Cassirer. Drawing, in particular, on Cassirer's claim in *Determinism and Indeterminism* that, "The *a priori* . . . must be understood in a purely methodological sense," Pap claims that, "Cassirer tends to assimilate Kant's doctrine of the *a priori* to the functional-pragmatic interpretation of the *a priori*, its interpretation as a methodological rule" (DI 74/91, 1943a). On Pap's reading of this passage, Cassirer indicates that the *a priori* should be understood in functional-pragmatist terms, as a "conceptual tool" we develop given our methodological needs, needs which can change as our theories do (1943a: 458). There are questions that could be raised over this pragmatist reading of Cassirer, especially given his critical remarks about pragmatism in *Substance and Function* (see SF 317–320/343–347); nevertheless, Pap's reading points toward an important instance of the early use made of Cassirer's dynamic account of the *a priori* in analytic philosophy of science.

More recently, Cassirer's account of the *a priori* has received attention in light of the broader revival of interest in Neo-Kantian themes in philosophy of science. We see this, in particular, in the work of philosophers of science, like Michael Friedman (2000), Alan Richardson (1998), Thomas Ryckman (2005), and André Carus (2007), who have paid close attention to early 20th century Neo-Kantian approaches to the *a priori* and its influence on logical positivists, like Carnap.

As I emphasized in Chapter Four, central to this discussion has been what kind of *a priori* Cassirer defends. This question has been raised, in one vein, as to where Cassirer's view stands in relation to Hans Reichenbach's view in *The Theory of Relativity and A Priori Knowledge* (1920 [1965]).[7] Reichenbach, who had been a student of Cassirer's in Berlin, draws a distinction between the 'absolute' *a priori* that is fixed for all time and the 'relativized' *a priori* that changes from theory to theory. And there has been much debate about whether Cassirer's views anticipate the latter conception of the relativized *a priori*. Cassirer's claim, for example, in volume one of *The Problem of*

Cognition (1906), to the effect that in science we "find only a relative stopping point, that we therefore have to treat the categories . . . as variable and capable of change, is obviously correct," seems to point in this direction (ECW 2, 12–13). However, given that in both his philosophy of science and philosophy of culture, Cassirer defends a theory of the categories as a priori concepts that have a fixed logical form, it seems there is room for an absolute kind of a priori on his view as well.

Meanwhile, in a second vein, it is debated whether Cassirer conceives of the a priori in constitutive or relative terms, that is, whether he takes the a priori to be constitutive of objects of scientific experience or whether it merely regulates the progress of scientific experience. While some commentators attribute to Cassirer only a regulative account of the a priori (Friedman, Ferrari), others allow for both the regulative and constitutive a priori to play a role on his view (Ryckman, Heis).[8]

In Chapter Four, I argued that attending to Cassirer's methodological remarks about natural science accommodates a view of the a priori that has absolute and relativized, constitutive and regulative elements. Yet regardless of which view of the a priori we ultimately take Cassirer to defend, his dynamic approach to the a priori has been yet another feature of his philosophy of science that has continued to have a lasting influence.

Cassirer and social science

Cassirer's philosophy of science has not only been influential for philosophers of science; it has also had an impact on social scientists. In particular, I shall highlight the influence Cassirer's approach to science had on two of his contemporaries, Kurt Goldstein, a pioneer of neuropsychology and Cassirer's cousin, and Kurt Lewin, the founder of modern social psychology.[9]

It is worth prefacing this discussion by noting that Cassirer himself engaged extensively with psychological research, particularly research on psychopathology and language by Henry Head, Adhemar Gelb, Ludwig Binswanger, and Goldstein, among others. The seriousness of his interest is evident in Chapter Six of the third volume of The Philosophy of Symbolic Forms, "On the Pathology of Symbolic

Consciousness," in which he analyzes pathological disorders, like aphasia, psychic blindness, and apraxia from the perspective of his philosophy of symbolic forms, and in "The Concept of Group and the Theory of Perception" (1938) in which he applies group theory to the psychology of perception.

Cassirer's own relation to psychology aside, as we shall see in what follows, for Goldstein and Lewin, what was most valuable in Cassirer's work for psychology and social science was his functional approach to science.[10]

Goldstein is perhaps best known for his work, The Organism (1934 [1995]). In this text, Goldstein appropriates Cassirer's analysis of the symbolic nature of physics in his holistic approach to psychology and biology. There, Goldstein applauds Cassirer for recognizing that natural science is not a matter of simply describing the facts that are given to us, but rather involves the functional transformation of what is perceived in light of scientific symbols and concepts. To this end, Goldstein quotes Cassirer's claim from The Philosophy of Symbolic Forms that in physics,

> Instead of the concrete data, we use symbolic images, which are supposed to correspond to data on the basis of theoretical postulates which the observer considers as true and valid. . . . The significance of these concepts is not manifest in the immediate perception, but can be determined and secured only by an extremely complex process of intellectual interpretation.
>
> (Goldstein 1995: 314, a translation of PSFv3 24/24/21)

Goldstein goes on to assert, "The type of biological knowledge, which we here advance, agrees in its fundamental tendency with the above-characterized epistemological approach" (314).[11] And Goldstein develops his holistic approach to biology and psychology on the basis of this Cassirer-style recognition of the functional role of symbols and the influence of theory on fact.

Cassirer's functional theory of science also played a significant role in relation to Lewin's pioneering work in social psychology.[12] As a first-year graduate student, Lewin took one of Cassirer's classes in Berlin in 1910, and his turn to psychology appears to have been spurred in part by Cassirer's comments on one of his papers, which questioned whether Lewin's philosophical claims were psychologically tenable (Marrow 1969: 9). And over the course of his career, Lewin expressed a debt of gratitude to Cassirer. To cite one example, in his piece on Cassirer for The Library of Living Philosophers, "Cassirer's Philosophy of Science and the Social Science" (1949), he says that from 1910 to 1946, "scarcely a year passed when I did not have specific reason to acknowledge the help which Cassirer's view on the nature of science and research offered" (272). And, in these remarks, Lewin makes clear that it was Cassirer's "analysis of the methodology and concept formation of the natural sciences" in Substance and Function that he found particularly promising for psychology in particular and social science in general (272). To illustrate Cassirer's significance for Lewin, I shall briefly canvas Lewin's account of Cassirer's relevance to social science and psychology.

With respect to social psychology, Lewin highlights the importance of Cassirer's account of the relationship between fact and theory as a relationship that involves not copying, but rather one in which theory helps determine facts. Applying this idea to social science, Lewin claims that the facts that pertain to social groups are not simply given; they are determined, in part, through concepts and laws that govern those social groups, which it is the responsibility of the social scientist to ascertain (Lewin 1949: 280–286).

Moreover, Lewin thinks that Cassirer's account of the centrality of quantitative analysis and the mathematization of nature in physics is something that has purchase in the social sciences. More specifically, he claims that the social sciences need to balance qualitative analysis with quantitative analysis. By his lights, quantitative analysis not only provides a more exact account of particular phenomena, but also paves the way for the integration of different fields across the social sciences (Lewin 1949: 286–288).

Finally, Lewin highlights the value of Cassirer's views with respect to developing a "systematic comparative theory of the sciences" (1949: 278). What Lewin has in mind is Cassirer's comparative analysis of fields, like "mathematics, physics, and chemistry," which brings to light the similarities of the procedures and conceptual problems across these fields, while at the same time preserving what is distinctive about each (276). Lewin thinks that integrating the social sciences into this kind of comparative practice would help clarify its place within the broader field of science.

Meanwhile, with respect to psychology, Lewin employs many of the features of Cassirer's view in formulating his distinctive methodology for psychology. Lewin makes this clear in his 1931 paper, originally in *Erkenntnis*, "The Conflict Between Aristotelian and Galilean Modes of Thought in Contemporary Psychology." There, Lewin argues that in order to make progress in psychology, what is needed is a methodological transition similar to what we see in physics from an Aristotelian to a Galilean model. And in defending this line of thought, he explicitly draws on Cassirer's analysis of the transition from substance- to function- modes of thinking in *Substance and Function* (see, e.g., Lewin 1931: 144, 149, 159, 169). By Lewin's lights, a transition to the Galilean view shifts our focus toward the functional laws that govern phenomena. And, according to Lewin, this not only allows for a more precise mathematical account of particular phenomena, but also paves the way for an integrated analysis of the relations of phenomena across different fields, for example, of the continuity between 'normal' and 'pathological' phenomena, or the continuity between 'optical' and 'auditory' phenomena (Lewin 1931: 159). This more systematic, functional approach is one that Lewin then develops in his own field-theoretical approach to psychology, which involves a functional analysis of psychological behavior in terms of the field forces that govern the whole situation and which is summed up in his behavior equation, according to which behavior (B) is a function of a person (P) and her environment (E): $B = f(P, E)$ (see Lewin 1931: 167).

In both the cases of Lewin and Goldstein, Cassirer's account of the functional nature and methodology of natural science serves as guide to the development of a functional methodology in biology, psychology, and social science more broadly, which does not take the facts as given, but rather takes them to be constructed in light of functional concepts.

Cassirer and intellectual history

Although from the start of his career, Cassirer was interested in developing his philosophy of science, he was also dedicated to doing justice to the historical development of ideas. And these writings on the history of ideas shaped two important trends in intellectual history in the 20th century.

To begin, Cassirer's historical analysis of the development of cognition from the Renaissance to the 20th century in the four volumes of *Das Erkenntnisproblem* (*The Problem of Knowledge* or *The Problem of Cognition*) were seminal for the burgeoning field of the history of science. These works were particularly influential for historians of science, including Emil Meyerson, Léon Brunschvicg, Helen Metzger, Edwin Burtt, Alexander Koyré, Anneliese Maier, and E.J. Dijksterhuis, many of whom were, in turn, influential for Kuhn.[13]

Furthermore, Cassirer's writings on the Renaissance and Enlightenment, including most famously *The Philosophy of Enlightenment* (1932), as well as *The Individual and the Cosmos* (1927), *The Question of Jean-Jacques Rousseau* (1932), and *The Platonic Renaissance in England* (1932), became touchstones in intellectual history for generations to come. For many, what was particularly significant about these works was Cassirer's holistic approach, that is, his effort to uncover the underlying intellectual trends in each epoch that unified together developments in fields ranging from philosophy to literature, art to science. Indeed, the value of this approach has been widely acknowledged by thinkers as distinct as Mikhail Bakhtin (1895–1975),[14] Paul Oskar Kristeller (1905–1999),[15] Michel Foucault (1926–1984),[16] and Peter Gay (1923–2015).

Cassirer's relation to Gay, in particular, is of note as Gay not only wrote the foreword to *The Philosophy of Enlightenment* and translated *The Question of Jean Jacques Rousseau*, but also offers a helpful analysis of Cassirer's value with respect to intellectual history in "The Social History of Ideas: Ernst Cassirer and After" (1967). In this piece, Gay assesses what he takes to be both the virtues and vices of Cassirer's approach to intellectual history. According to Gay, the merits of Cassirer's approach are threefold. To begin, he claims that Cassirer's method is one that resists treating history as either art or science, and instead treats it as both, demanding that the historian be at once objective and imaginative (Gay 1967: 108). Second, Gay maintains that Cassirer's view of the "primacy of function" becomes a powerful tool for analysis in intellectual history insofar as it allows one to investigate the function and force of ideas as drivers of history

(110). And, finally, Gay says that Cassirer is a "supreme example" of how to approach intellectual history from a 'synthetic' perspective, by looking not for contradictions, but rather for the coherence in historical thinkers and periods (117).

This being said, Gay nevertheless faults Cassirer's methodology as follows: "The really serious difficulty in Cassirer's conception of intellectual history is . . . his failure to do justice to the social dimensions of ideas" (1967: 117). By Gay's lights, Cassirer approaches the development of ideas in history in isolation from the social forces that shape them and, in this regard, Gay thinks Cassirer's method remains incomplete. Criticisms aside, however, Gay nevertheless calls our attention to Cassirer's pivotal contribution to the development of intellectual history in the 20th century.

Cassirer's *Philosophy of Symbolic Forms* and the semantic turn

Having considered the legacy of Cassirer's philosophy of science and historical work, I shall now turn to the reception of his philosophy of symbolic forms. Some regard this aspect of Cassirer's philosophy as the most groundbreaking one insofar as it contributes to the so-called 'semantic turn' in 20th century thought, which places symbols, signs, and meaning at the center of human and cultural activity. Indeed, for thinkers in fields ranging from philosophy to anthropology, art history to linguistics, Cassirer's *Philosophy of Symbolic Forms* represents one of the first serious philosophical attempts to work out a theory of symbolism and, as such, it serves as the starting point for them to develop their own semantic frameworks.

In order to bring to light the ways in which Cassirer's basic symbolic picture influences thinkers working in different fields, in this section I begin by exploring the impact of Cassirer's general theory of symbolism on the American philosophers, Susanne Langer and Nelson Goodman, and on the French phenomenologist, Maurice Merleau-Ponty. I then shift my focus to his impact on the semantic approach to more specific regions in culture, including art, myth, and language.

With regard to the influence of Cassirer on Langer, Goodman, and Merleau-Ponty, as we shall see, in spite of the differences between these philosophers, what draws them all to Cassirer's theory of symbolism is the way in which his approach allows for a non-dualistic account of the continuity and diversity of human experience and culture. Unlike dualistic models, which force us to divide activities into those that belong either on the side of nature or reason, Cassirer's symbolic model enables us to recognize a wide-range of activities as belonging on the same plane insofar as they are symptomatic of our nature as an "*animal symbolicum*" (EM 26/31). If we follow Cassirer's account, at the level of individual human experience, we no longer need to draw a distinction between those activities, like perception or emotion, which might seem to fall outside the 'space of reasons', and those like language-use and cognition, that fall within it. Instead, in keeping with Cassirer, we can situate all of these activities on a symbolic continuum, as the various ways in which individuals try and symbolically make sense of our selves, others, and the world. Meanwhile, when it comes to culture, rather than designating certain cultural domains, like myth or art, as irrational, and others, like language or science, as rational, if we take our cue from Cassirer, then we can recognize them as of a piece, as large-scale efforts to symbolically construct the world. And for Langer, Goodman, and Merleau-Ponty it is this semantically-oriented account of the unity of human experience and culture, which drew them to Cassirer.

Cassirer and Langer

Arguably, Langer's philosophy represents the most extended attempt on offer to further develop Cassirer's philosophy of symbolism. Langer was a wide-ranging philosopher, best known for her work in symbolic logic, philosophy of mind, and aesthetics. Though her early works, like Practice of Philosophy (1930) and Introduction to Symbolic Logic (1937), betray the influence of her teacher, Alfred North Whitehead, her study of Cassirer's philosophy in the 1930s and her personal interactions with him in the U.S. shaped her subsequent works in significant ways.[17] Cassirer's influence first becomes evident in her perhaps best known work, Philosophy in a New Key (1942 [1957a]), and continues to shape her subsequent writings, including Feeling and Form (1953) (which she dedicates to Cassirer), Problems of Art (1957b), and her three-volume work Mind (1967,

1972, 1982). In addition to these writings, she also translated Cassirer's *Language and Myth* into English in 1946.

One of the central reasons Langer gravitates towards Cassirer's philosophy of symbolic forms is because she regards it as one of the catalysts for philosophy's modulation in the 20th century into a 'new key', the 'keynote' of which is "the fundamental notion of symbolism" (1957a: 25). For her part, Langer concentrates much of her energy on developing a philosophy of mind and aesthetics that takes its cue from this key change in general and Cassirer's version of it in particular.

In her philosophy of mind, Langer takes as her point of departure the idea that "symbolism is the key to that mental life which is characteristically human and above the level of sheer animality" (1957a: 28). In this, Langer's view closely parallels Cassirer's account of human beings as 'symbolic animals' and his analysis of the human mind in terms of different modes of symbolic understanding. Moreover, Langer makes clear that symbolic activities of the human being cannot be understood in passive terms, as a matter of us symbolically copying the world around us; rather she argues that symbolism involves the activity of 'transformation' by means of which we transform our current experiences into something meaningful (see 1957a: Ch. 2). On this point too, Langer's view parallels Cassirer, with his rejection of the copy theory of cognition in favor of an account of the active and transformative symbolic activities involved in our understanding of the world.[18]

Langer further draws on Cassirer's distinction between 'discursive' and 'non-discursive' symbols in order to fill out the details of her account of the mind. According to her, there is a tendency in the philosophy of mind to analyze the mind only in terms of its *discursive* capacities and activities, for example, those involved in language, mathematics, or science. Though she thinks that these discursive activities account for part of our symbolic activities, she, following Cassirer, argues that there are also non-discursive activities that we engage in, for example, in myth and art, that are no less symbolic, hence no less valuable for the analysis of the mind than our discursive activities (see 1957a: Ch. 4).[19] Ultimately, Langer thinks Cassirer's recognition of the distinction between discursive and non-discursive symbols points toward the need to broaden our approach of philosophy of mind. In addition to accounting for the discursive symbolism we engage in through reason, she thinks we must also account for the non-discursive symbolism we engage in through other modes of understanding, like intuition and imagination.[20] It is only on the basis of this type of broader account of symbols and symbolic understanding that Langer, like Cassirer, thinks we can give a full account of the human mind. And it is for this reason that Langer's work in the philosophy of mind includes extensive analyses of language, myth, and, above all, art: she sees this detailed cultural analysis as crucial for bringing to light the continuity between the discursive and non-discursive symbolic activities of the mind.

As for Langer's philosophy of art itself, she draws on Cassirer in defending a cognitivist approach to aesthetics.[21] As we saw in Chapter Six, Cassirer's view of art is cognitivist in the sense that he allows for art to give us knowledge of 'intuitive' forms, like the forms of feelings and spatial and temporal forms of objects out-side of us. Though Langer largely sets aside questions regarding art's relation to external things, she follows Cassirer's lead in arguing that art gives us cognition of human feelings by means of its non-discursive symbols (see, e.g., Langer 1953: 28, 40; 1957b: 15, 25).[22] Illustrating this point initially in the context of music (1957a: Ch. 8), she then extends it to an analysis of the distinctive ways in which various artistic media give us this kind of knowledge of feeling in *Feeling and Form* and *Problems of Art*.

Though in these ways Langer's philosophy of mind and aesthetics take their cue from Cassirer's philosophy of symbolic forms, there are two significant points on which Langer appears to part ways with Cassirer. To begin, unlike Cassirer who situates his account of sym-bolism in the framework of critical idealism, Langer strives to offer an account of symbolism that remains neutral with regard to onto-logical issues (see 1957a: xiv). Moreover, whereas Cassirer traces our symbolic activities back to the self-activity of spirit and the spon-taneity of human beings, Langer offers a much more naturalized account of symbolism. As she makes clear already in *Philosophy in a New Key*, she thinks of symbolic activities as the result of "primary needs" that arise organically in human beings (1957a: 38). And explicating this organic basis becomes the primary task of her last major work, *Mind*. Differences aside, however, it is clear that Langer's philosophy of mind and symbolism more generally embodies her effort to carry forward Cassirer's *Philosophy of Symbolic Forms*.

Cassirer and Goodman

The second American philosopher whose thought was impacted in significant ways by Cassirer's philosophical symbolism is Nelson Goodman. Goodman was one of the most influential American philosophers in the post-war period, whose wide-ranging interests included logic, metaphysics, epistemology, semantics, philosophy of science, and philosophy of art. However, Cassirer's impact on his thought is most evident in two areas: his theory of symbols and aesthetics. Though the influence of Cassirer is more elliptical in the case of Goodman than Langer, in *Languages of Art* (1968 [1976]) and *Ways of Worldmaking* (1978) Goodman takes his cue from Cassirer's insight into the need for a broad, systematic account of symbolism, which can serve as the key to elucidating the different modes of human understanding in general and those involved in art, in particular.

In the introduction to *Languages of Art*, Goodman indicates that the subtitle of the book, *An Approach to a Theory of Symbols*, better articulates his aim, viz., to offer a "general theory of symbols" (1976: xi). Goodman claims that a general theory of symbols is needed because a "systematic inquiry into the varieties and functions of symbols" is one that philosophy tends to neglect and he approvingly cites Cassirer and Langer for attempting to remedy this lacuna (1976: xi–xii). Goodman thus situates his work in relation to the broader project of philosophy of symbolism and treats Cassirer and Langer as allies to this end.

Though committed to the same project, Goodman nevertheless notes later in *Ways of Worldmaking* that he and Cassirer approach a theory of symbols from two different angles. Whereas Cassirer's approach is oriented around "a cross-cultural study of the development of myth, religion, language, art, and science," Goodman maintains that his "approach is rather through an analytic study of types and functions of symbols and symbol systems" (1978: 5). Though by no means exclusive, Goodman thus regards his more formal analysis of the logic of symbols to be approaching the philosophy of symbols in a different direction from Cassirer's culture-based approach.

This difference aside, Goodman's philosophy of symbolism parallels Cassirer's in, at least, three ways. To begin, Goodman, like Cassirer, insists on a recognition of different modes of symbolism, and this culminates in Goodman's discussion of distinctive kinds of symbolism: denotation, exemplification, and expression. Moreover, according to Goodman, a symbol can only be understood as part of a broader symbolic 'system'. And in this regard, Goodman's view displays affinities with Cassirer's commitment to situating an account of individual cultural configurations in the systematic context of a symbolic form.

Finally, for Goodman, as for Cassirer, symbols have a "formative function": instead of copying the world, they construct or 'form' it (1978: 1). This last point is one that Goodman addresses in the first chapter of *Ways of Worldmaking*, where he offers an interpretation of Cassirer's idealism. There, Goodman claims that he and Cassirer share a basic commitment to rejecting the idea that the world is simply 'given' to us, in favor of a more constructive view, according to which the world is constructed through the creative acts of the understanding and the formative power of symbols. Goodman, moreover, asserts that he and Cassirer share a pluralistic account of idealism, according to which, "the multiplicity of worlds, the speciousness of 'the given', the creative power of the understanding, the variety and formative functions of symbols" (1978: 1). On this pluralistic view, there are a plurality of worlds that correspond to each symbol system. There is a question of whether this is a view Cassirer, in fact, defends. Though he, at times, indicates that there is a world for each symbolic form (see, e.g., PSFv1 18/18/88), as we saw in Chapter Five, he is also committed to each symbolic form being a part of a single common world of humanity or symbolic universe.

For Goodman, one of the most significant symbol systems is that of art and, on this topic, we find that he defends a cognitivist approach to aesthetics that has affinities with the cognitivist aesthetics of Cassirer and Langer.[23] For, like Cassirer and Langer, Goodman argues that "aesthetic experience is cognitive experience" in the sense that the symbols involved in art give us understanding (1976: 262). For example, on Goodman's view the denotative symbols involved in pictures give us knowledge of what they denote, while expressive symbols give us understanding of what is metaphorically exemplified in the work of art. In the cognitivist spirit of Cassirer's aesthetics, then, Goodman acknowledges the ways in which the symbol system of art can give us a distinctive kind of knowledge.

Cassirer and Merleau-Ponty

Shifting away from the American philosophical tradition toward the phenomenological one, although much attention has been paid to the disputation between Cassirer and Heidegger,[24] less has been paid to Cassirer's influence on Merleau-Ponty.[25] However, as citations throughout the Phenomenology of Perception (1945 [2013]) reveal, the third volume of Cassirer's Philosophy of Symbolic Forms was a significant resource that Merleau-Ponty drew on in working out his theory of perception and culture.[26] This may perhaps come as a surprise since throughout the Phenomenology of Perception, Merleau-Ponty pits his body-centric account of perception against an 'intellectualist' Kantian account of perception, and specifically targets Cassirer in doing so (see Merleau-Ponty 2013: 522, fn. 67). Nevertheless, Merleau-Ponty also claims that Cassirer's view contains certain "phenomenological and even existential analyses" that he intends to "make use" of in his own theory of perception and culture (522, fn. 67).

More specifically, there are two aspects of Cassirer's account of symbolic forms that Merleau-Ponty gravitates towards: Cassirer's account of the relationship between the expressive, presentative, and significative functions of consciousness, and his conception of 'symbolic pregnance' (see Chapter Five). Merleau-Ponty highlights these aspects of Cassirer's philosophy because he thinks they reveal that in addition to the more discursive and intellectual ways we have of experiencing meaning in the world through the presentative and significative functions, there is a more primordial experience of meaning we have as mediated through the function of expression. In a Cassirerean vein, in the Phenomenology, Merleau-Ponty says,

> We must acknowledge "expressive experiences" (Ausdruckerlebnisse) as prior to "acts of signification" (bedeutungsgebende Akten) by theoretical and thetic consciousness; we must acknowledge "expressive

> sense" (*Ausdrucks-Sinn*) as prior to "significative sense" (*Zeichen-Sinn*):
> and we must acknowledge the symbolic "pregnance" of form in
> content prior to the subsumption of content under form."
>
> (Merleau-Ponty 2013: 304)

As we see in this passage, Merleau-Ponty thinks that we need to
recognize that in addition to significative acts of consciousness in
which we discursively subsume particular contents under conceptual forms, there are expressive experiences in which we experience
contents that are already 'pregnant' with form.[27]

Merleau-Ponty, in turn, argues that these expressive experiences
have their locus in the body:

> My body is the place, or rather the very actuality of the phe-
> nomenon of expression (*Ausdruck*); in my body, visual and audi-
> tory experience, for example, are pregnant with each other,
> and their expressive value grounds the pre-predicative unity
> of the perceived world, and through this, its verbal expres-
> sion [*l'expression verbale*] (*Darstellung*) and intellectual signification
> (*Bedeutung*).
>
> (Merleau-Ponty 2013: 244)

According to Merleau-Ponty, our bodies enable us to understand the
world in an expressive and pregnant way, which occurs prior to discursive acts of presentation or signification. Moreover, he argues that
these expressive embodied experiences are what enable us to grasp
the cultural world. To this end, in the *Phenomenology*, he claims that the
body grounds the expressive understanding involved in myth, the
presentative grasp we have through language, and the significative
grasp we have through the intellect. And in his unfinished manuscript, *The Prose of the World*, he maintains that bodily expression and
symbolic pregnance play a key role in our engagement in art, mathematics, and natural science.

In these ways, though Merleau-Ponty is critical of the residual
intellectualist features of Cassirer's philosophy of symbolism, he
nevertheless regards Cassirer's account as an important resource to
draw on in defending his account of perception and culture. And
though Merleau-Ponty comes from a different tradition than Langer

and Goodman, we find that, for all three, Cassirer's general philosophy of symbolism is invaluable insofar as it sheds light on the variety of discursive and non-discursive symbols that give human experience and culture its richness and depth.

Cassirer, Panofsky, and art history

With respect to the study of art, Cassirer's influence extends not just to the cognitivist aesthetics of Langer and Goodman, but also to one of the most influential figures in art history in the 20th century, Erwin Panofsky.[28] Panofsky is best known for his contributions to the development of iconography as a distinctive branch of art history. And, as we shall see, his conception of the iconographical method was informed in important ways by his engagement with Cassirer in Hamburg at the Warburg Library in the 1920s.[29] In order to explore Cassirer's influence on Panofsky I begin with a discussion of the iconographical method, in general, and its connection to the notion of a 'symbolic form', before looking at how Panofsky uses this method with respect to perspective in *Perspective as Symbolic Form* (1924–1925/1927).[30]

Panofsky presents the iconographical method as a new method for art history, which differs from the more formalist method offered by Heinrich Wölfflin and the more contextualist method offered by Alois Riegl. By Panofsky's lights, the problem with Wölfflin's purely formal approach is that it neglects the semantic elements of a work of art. Panofsky is more sympathetic to Riegl's contextualist approach, which seeks to understand the "unity of the creative forces—forces both of form and content—which organize the work from within" (Panofsky 1981: 18). From Panofsky's perspective, in clarifying these 'creative forces' Riegl's method does better justice to the semantic aspects of works of art (1981: 18). However, Panofsky ultimately criticizes Riegl's view as too psychologistic (1981: 18). According to Riegl, the key to understanding the unity of form and content is the notion of 'Kunstwollen', 'artistic volition' or 'artistic will'. And, at least on Panofsky's reading, Riegl conceives of Kunstwollen in a psychologistic way, that is, in terms of the psychology of an artist, historical period, or appreciator (see Panofsky 1981: §§1–3). On Panofsky's view, this is a mistake: the ultimate source of the unity of form and content of a work of art is not a set of extrinsic psychological facts, but rather something intrinsic to the work of art that unifies form and content from within, which he calls its "immanent sense" (Panofsky 2008: 44).

Though Panofsky had been working out many of these ideas prior to his appointment at the University of Hamburg in 1920,[31] he found in Cassirer's account of symbolic forms a more precise way of articulating the immanent sense of a work of art, viz., in terms of the symbolic forces that unify the form and content of a 'symbolically pregnant' work of art from within. Thus, by recasting Riegl's notion of *Kunstwollen* in terms of Cassirer's notion of symbolic form, Panofsky was able to retain Riegl's insight that there is a 'unity of creative forces . . . that shape the work from within', while grounding it in an account of the symbolic forms of culture that resists the psychologistic trappings of Riegl's view.

Panofsky explicitly emphasizes the importance of Cassirer's thought for the development of his iconographical method in the Introduction to *Studies in Iconology*. There he distinguishes between three levels of an analysis an art historian can pursue. First, there is 'pre-iconographical' analysis, which aims at interpreting the 'primary' or 'natural' meanings that are connected to everyday objects, events, and emotions and that are expressed through artistic 'motifs' (Panofsky 1972: 14, 5). Second is "*iconographical analysis in the narrower sense*," which concentrates on 'secondary' or 'conventional' meanings, that is, "themes or concepts" articulated through "images, stories, and allegories" (14, 6). Finally, there is "*iconographical interpretation in a deeper sense*," which concerns 'intrinsic' meanings, that is, those that pertain to the "'*symbolical*' values" of a work of art that express the "underlying principles which reveal the basic attitude of a nation, a period, a class, a religious or philosophical persuasion" (14, 8–9). As Panofsky makes clear, his understanding of the latter form of iconography takes its cue from Cassirer: "conceiving of pure forms, motifs, images, stories and allegories as manifestations of underlying principles, we interpret these elements as what Ernst Cassirer has called 'symbolical' values" (8). Panofsky's idea is that by analyzing the forms, motifs, images, stories, and allegories involved in works of art along Cassirer's symbolic lines, we seek to understand the way in which the underlying cultural principles symbolically shape the work of art within and give it its immanent sense.

Panofsky pursues this Cassirer-inspired form of iconography in *Perspective as Symbolic Form*. In this essay, Panofsky introduces the idea of perspective as a symbolic form in the context of his account of the development of perspective from antiquity to the Renaissance. In order to explain this development, Panofsky argues that perspective should not be regarded as something simply given through perception; it should, instead, be understood as something that symbolically expresses the "sense of space" and "sense of the world" that is indexed to each epoch (Panofsky 1991: 34). For Panofsky, perspective is thus symbolic of the sense of space and the world, which animates a unique cultural period and which is immanent to a work of art. And it is for this reason that Panofsky ultimately describes perspective as a symbolic form:

> perspective . . . may even be characterized as (to extend Ernst Cassirer's felicitous term to the history of art) one of those "symbolic forms" in which "spiritual meaning is attached to a concrete, material sign and intrinsically given to this sign." This is why it is essential to ask of artistic periods and regions not only whether they have perspective, but also which perspective they have
>
> (Panofsky 1991: 40–41).[32]

As we see here, Panofsky describes perspective as a symbolic form because it involves an immanent sense that unifies the form and content of a work of art from within. To be sure, this is something of an extension of Cassirer's view, since it seems Cassirer himself would regard art as a symbolic form, which perspective is a moment of. Nevertheless, Panofsky deploys the concept of the symbolic form in this way because he finds it fruitful with respect to the iconographical aim of articulating the immanent sense that propels the development of perspective in art. And in this way *Perspective as Symbolic Form* exemplifies the broader use Panofsky makes of Cassirer's philosophy of symbolic forms in developing and executing his iconographical method.

Cassirer, Blumenberg, and myth

While Cassirer's symbolic approach to myth displays affinities with some of the major trends in anthropology in the 20th century, like

Clifford Geertz's symbolic anthropology and Claude Lévi-Strauss's structural anthropology,[33] it is the German philosopher and intellectual historian, Hans Blumenberg whom Cassirer influences most directly.[34]

Blumenberg is best known for his work on metaphor, myth, and the 'modern age', and though Cassirer's account of myth had a significant impact on Blumenberg, he also has an ambivalent attitude towards Cassirer. On the one hand, he is sympathetic to Cassirer's general emphasis on the symbolic nature of human existence. And, for Blumenberg, our symbolic nature manifests in a central way through the pervasiveness of metaphor. Moreover, Blumenberg thinks that Cassirer's account of myth takes steps in an important direction insofar as it treats myth not just as something that is irrational or illogical, but rather as something that has its own kind of logic, distinct from the theoretical logic of science (see, e.g., Blumenberg 1985: 362).

On the other hand, Blumenberg criticizes Cassirer's account of culture and myth on three fronts. In the first place, Blumenberg rejects Cassirer's teleological analysis of the relationship of the symbolic forms. Blumenberg raises this objection in "Ernst Cassirer Remembered" ("Ernst Cassirers Gedenkend") (1974), his acceptance speech for the Kuno Fischer prize, which Cassirer himself had won in 1914. There he criticizes Cassirer for his teleological commitment to natural science as the stage of culture that the other symbolic forms, in some sense, lead towards. According to Blumenberg, not only is this teleological view in tension with Cassirer's other commitment to the irreducibility of all the symbolic forms, but also that it presupposes a problematic view of history as having as its goal the present state of science. For these reasons, Blumenberg suggests that, "What remains to be learned from Cassirer is to be found in what he was unable to accomplish," viz., in a historical and symbolic analysis of culture that eschews teleological trappings (1974: 460).[35]

Second, Blumenberg objects to Cassirer's account of the origin of human beings as symbolic animals. On this point he appears to be persuaded by Heidegger's criticisms of Cassirer at Davos, which Blumenberg had studied in the 1970s.[36] In particular, Blumenberg draws on Heidegger's critique of Cassirer with regard to the question of the *terminus ad quem* and *terminus a quo* of culture (see Heidegger

1990: 202). According to Heidegger, Cassirer approaches culture from the perspective of the 'terminus ad quem', that is, from the perspective of the goal of freedom towards which culture is working; however, he charges that as a result of adopting this perspective, "the terminus a quo is utterly problematical" for Cassirer (202). That is to say, Heidegger thinks Cassirer's Philosophy of Symbolic Forms fails to take into account the origin from which our cultural activities spring, a project that Heidegger believes requires a "metaphysics of Dasein" and an analysis of finitude and thrownness (202).

Taking over this criticism in "An Anthropological Approach to the Contemporary Significance of Rhetoric" ("Anthropologische Annäherung an die Aktualität der Rhetorik") (1971 [2001]), Blumenberg objects that Cassirer simply assumes that human beings are symbolic animals and fails to address the more fundamental terminus a quo question regarding the source of this symbolic activity (see, e.g., 438). Continuing in a Heideggerean vein, Blumenberg then argues that we develop our symbolic capacities in response to the anxiety that we feel as organisms that are confronted with a reality that we are contingently subject to. Symbolic activities, on Blumenberg's view, should thus be understood as our attempt to gain control of our situation by distancing ourselves from this reality.

According to Blumenberg, Cassirer's teleological assumptions about culture and his failure to offer an account of the terminus a quo of us as symbolic animals, in turn, give rise to a problematic account of myth, and this is the third area in which he criticizes Cassirer. To this end, in Work on Myth (1979 [1985]), Blumenberg argues that by making the teleological assumption that science is a kind of terminus ad quem of culture, Cassirer's view renders myth 'obsolete' over time:

> The theory of the symbolic forms allows one for the first time to correlate the expressive means of myth with those of science, but in a historically irreversible relationship and with the unrelinquishable presupposition of science as the terminus ad quem. Myth is made obsolete by what comes after it; science cannot be made obsolete, however much, in each of its steps forward, it itself makes the preceding steps obsolete.
>
> (Blumenberg 1985: 50)

Not only does Blumenberg think that this position is in tension with Cassirer's attempts to secure the 'legitimacy' of myth, but also that it ignores the ways in which myth persists in culture (1985: 51). To be sure, he recognizes that in Myth of the State Cassirer attempts to address this issue; however, he thinks that this later view does not fit with the view of myth he defends in The Philosophy of Symbolic Forms. As an alternative view of myth, Blumenberg suggests we approach myth from the terminus a quo perspective:

> My opinion, in contrast to [Cassirer's], is that in order to per-
> ceive myth's genuine quality as an accomplishment one would
> have to describe it from the point of view of its terminus a quo.
> Removal away from, not approach toward, then becomes the cri-
> terion employed in the analysis of its function. . . . Understood
> in terms of . . . a means of self-preservation and stability in the
> world.
>
> (Blumenberg 1985: 168).

Thus, though Blumenberg is sympathetic to Cassirer's recognition of the symbolic nature of myth and of human existence more
generally, he thinks that we should reject Cassirer's teleological
approach in favor of a more existentially-informed, organic terminus
a quo approach. And in this attempt to naturalize Cassirer's account of
symbolic animals and symbolic forms, Blumenberg's project bears
important similarities to Langer's project in Mind.

Cassirer, the Bakhtin Circle, and language

In addition to his impact on the study of art and myth in the
20th century, Cassirer's views of language played an important
role for members of the Bakhtin School of literary criticism.[37] The
Bakhtin Circle was a group of Russian intellectuals who began
meeting after the Russian Revolution to discuss social, cultural,
and political matters, with a special interest in the role language,
both ordinary and literary, plays in these complex networks.
The group originally began by reading Kant, a move inspired
by Matvei Kagan, who had studied with Cohen and Cassirer. Of
the members of the group, however, it is Valentin Voloshinov and

Mikhail Bakhtin who appear to have been most directly influenced by Cassirer.[38]

Voloshinov was a linguist and philosopher of language who, along with Pavel Medvedev, helped orient the Bakhtin Circle towards issues in language and semiotics in the late 1920s. He is best known for his book *Marxism and the Philosophy of Language* (1929 [1986]) in which he defends a Marxist analysis of language as a socially constructed and historically dynamic semiotic system that is a medium of ideology. While he was preparing this book, he was also in the process of translating the first volume of *The Philosophy of Symbolic Forms* (a project he was not able to finish before his early death). And, as his 1928 working notes indicate, he takes Cassirer's philosophy of language seriously because it represents a promising alternative to abstract idealist accounts of language that leave no room for the dynamic and historic nature of language, on the one hand, and psychologistic accounts that treat language as a matter of individual psyches, on the other. Hence Voloshinov notes, for Cassirer,

> 'the word' becomes a *partition* between transcendental validity and concrete actuality, a 'third kingdom', as it were lying between the cognizing psycho-physical subject and the empirical actuality surrounding him on the one hand, and the world of *a priori*, formal being on the other. . . . It is precisely on the ground of the philosophy of language that the Marburg Schools' scientism and logicism . . . are presently being overcome. . . . By means of the inner forms of language (the semi-transcendental forms, as it were) movement and historical becoming are being introduced into the petrified kingdom of the transcendental-logical categories.
>
> (Voloshinov 2004: 232)

Moreover, insofar as Cassirer acknowledges the constructive and creative nature of language, Cassirer's philosophy of language fits with Voloshinov's own. This said, Voloshinov develops this account of language in a Marxist direction and introduces the idea that dialogue is the 'inner form' of language. And he resists Cassirer's idealism in favor of a more (Bühler-inspired) realist view, which treats language as a means of 'refracting' reality; nevertheless, he sees in Cassirer's approach a helpful starting point for developing his own view (Voloshinov 1986: 9).

Turning to Bakhtin, both Cassirer's historical works and his philosophy of symbolic forms were of significance for him. In the former vein, Bakhtin is generally sympathetic to Cassirer's historical analysis of the overarching worldview of the Middle Ages and Renaissance and Cassirer's synthetic method. He also draws on Cassirer's account of comedy as something that has a cognitive and liberating function.[39] This influence is evident in Rabelais and His World (1965, composed in the 1930s), in which Bakhtin cites and, indeed, plagiarizes passages from Cassirer's Individual and the Cosmos and The Platonic Renaissance in England.

Bakhtin also closely read the first two volumes of The Philosophy of Symbolic Forms and he drew on them in developing his mature account of literature, genre, and the novel. The influence of Cassirer's philosophy of symbolism is perhaps most evident in Bakhtin's draft of the chapter titled, "Forms of Time and of the Chronotope in the Novel" for his manuscript The Bildungsroman and its Significance in the History of Realism (1937–1938 [2007]). There, he extends Cassirer's analysis of how language 'assimilates' time in different ways to an analysis of how the novel assimilates time in different ways, for example, through 'castle' time or 'biographical' time (see Bakhtin 2007: 251, 246, 249). Though Bakhtin thus appropriates Cassirer's philosophy to more literary ends, he, like Voloshinov, sees in Cassirer an approach to language that treats it as an index of our social, historical, and cultural situation, a situation that is symbolically constructed through that very language.

Cassirer and critical theory

The final aspect of Cassirer's legacy that I shall consider pertains to his influence for critical theory and on the contemporary critical theorists, Jürgen Habermas, Drucilla Cornell, and Kenneth Michael Panfilio.

Habermas and the linguistic turn

Habermas is an influential German philosopher who belongs to the second wave of the Frankfurt School and who has made decisive contributions to social and political theory, epistemology, aesthetics, and the philosophy of language. In his essay, "The Liberating Power of Symbols: Ernst Cassirer's Humanistic Legacy and the Warburg Library" (1996), Habermas explores the significance Cassirer's philosophy of symbolism in general and his philosophy of language in particular had for the post-war, critical tradition in Germany.[40]

According to Habermas, Cassirer's "original achievement consists in a semiotic transformation of Kantian transcendental philosophy" (Habermas 2001: 12). As Habermas presents it, Cassirer's semiotic transformation involves a reorientation of Kantian philosophy away from an analysis of how a transcendental subject constitutes the world toward an analysis of how intersubjectively shared symbolic forms confer meaning on the world (see, e.g., 14–15). By Habermas's lights, this is an important achievement insofar as it overcomes the latent dualism in Kant's philosophy, which posits a rigid distinction between subjects and objects, in favor of a more holistic view that points toward a "symbolic medium [that] has a structure which embraces both the internal and external" (14). Moreover, Habermas suggests that Cassirer's view has the virtue of shifting away from a problematic metaphysical conception of the 'thing in itself' and recognizing, instead, that there is no 'given' that exists outside of the intersubjectively shared symbolic world (see 16–17).

From Habermas's perspective, what is of particular significance for critical theory is Cassirer's semiotic approach to language:

> Cassirer was the first to perceive the paradigmatic significance of Humboldt's philosophy of language; and he thus prepared the way for my generation, the post-war generation, to take up the 'linguistic turn' in analytical philosophy and integrate it with the native tradition of hermeneutic philosophy.
>
> (Habermas 2001: 12)

For Habermas, Cassirer's semiotic approach to language offered an alternative to nominalist theories of language. As Habermas presents Cassirer's alternative, although naming and reference are important, the most decisive achievement of language concerns its "disclosing function" (Habermas 2001: 13). This 'disclosing function' amounts to the ways in which the "intersubjectively shared domain" of language productively structures and articulates the world in such a way that first confers meaning on it (14). Articulating the broader significance of this account of language, Habermas claims,

> [Cassirer's] notion of language implies more than just a new linguistic theory. By commandeering Kant's notion of the transcendental, so to speak, and transforming the world-constituting activity of the knowing subject into the world-disclosing function of the trans-subjective form of language, it explodes the architectonic of the philosophy of consciousness as a whole.
> (Habermas 2001: 15)

On Habermas's reading, Cassirer's philosophy of language thus paved the way toward a critical theory of language by recognizing the world-disclosing function of language, qua something we intersubjectively share.

This being said, Habermas is critical of Cassirer's philosophy of language for not having gone far enough. Habermas introduces this objection in the context of discussing what he takes to be a fundamental tension between Cassirer's commitment to the equality and teleology of the symbolic forms (see Habermas 2001: §3). According to Habermas, Cassirer's commitment to the equality of the symbolic forms undermines the prospects for any teleological or unified account of them. Rather, he takes Cassirer to be forced to endorse a kind of 'pluralism' and 'perspectivism' (Habermas 2001: 20). However, Habermas objects that within this pluralistic framework, the symbolic forms are ultimately 'incommensurable' with one another and, as a result, Cassirer has no way to account for the "civilizing process" of culture and the "unity of reason" (24, 21). Instead, in Cassirer's system, "the identity of reason itself [is] dissolved into a multiplicity of contexts" (21). According to Habermas, the only way out of this situation is by taking a "step that Cassirer was unwilling to take," viz., giving language "systematic priority" over the other symbolic forms (22). Habermas defends his own version of this view with his communicative theory of rationality, and he thinks that such a view is needed in order to account for the unity of reason and the teleological emancipation that culture involves.

The latter topic points us toward the last feature of Habermas's reading of Cassirer, which I shall mention: the 'humanist legacy' of Cassirer's philosophy. According to Habermas, the humanist promise of Cassirer's philosophy of symbolism is connected to his recognition of symbolization as part of the process that "drives the process of civilization forward, . . . [and] promotes increasingly civilized behavior" (Habermas 2001: 25). Although Habermas thinks that the progress towards 'civilization' is something that Cassirer, with his epistemological perspective, does not always bring out, it is the cornerstone of his humanist legacy—a legacy that Habermas's philosophy attests to.

Cornell and Panfilio on the revolutionary power of symbolic forms

Whereas Habermas's analysis of Cassirer's humanist legacy highlights the value of his views with respect to the civilizing process in general, Drucilla Cornell and Kenneth Michael Panfilio focus on this legacy with respect to the concrete racial, social, and political problems stemming from the colonialism and capitalism that dominate modernity.[41] To this end, in *Symbolic Forms for a New Humanity: Cultural and Racial Reconfigurations of Critical Theory* (2010), they bring Cassirer's philosophy of symbolic forms into dialogue with Africana philosophers and critical theorists, like Frantz Fanon, Paget Henry, C.L.R. James, and Lewis Gordon. And they argue that out of this dialogue there emerges a form of "ethical humanism," which turns on a commitment to the revolutionary reconfiguration of problematic structures through symbolic means (Cornell and Panfilio 2010: x).

In order to bring out the relevance of Cassirer's views to these critical issues, Cornell and Panfilio argue that we must resist the tendency to read Cassirer, as is often done, as a "facile, optimistic humanist" (2010: 3). Far from being facile or overly optimistic, they argue that Cassirer's humanism is one that recognizes the work, struggle, and revolution required for culture, on the one hand, and the genuine threat to the ethical ideals of culture from politically-entrenched racism, on the other. Accordingly, they devote much of

Symbolic Forms for a New Humanity to a careful reading of *The Philosophy of Symbolic Forms, Language and Myth, An Essay on Man*, and *The Myth of the State*, with the intention of bringing out the subtlety of Cassirer's humanism and ongoing relevance.

More specifically, on their interpretation, Cassirer's ethical humanism turns on two key claims. First, they attribute to Cassirer a pluralistic account of the symbolic forms, according to which, "there is no clear hierarchy of symbolic forms and, as a result, no obvious hierarchy of human beings who represent the highest form of reason and the highest form of thinking" (Cornell and Panfilio 2010: 12). Second, they argue that, for Cassirer, the symbolizing process involved in the symbolic forms is intersubjective in character: "we come to the world in a symbolic universe where symbolization is not reducible to a subjective capacity; and as we enlarge our perspective about who we are our humanity is enhanced" (12).

By their lights, appreciating these two claims together points towards the ways in which Cassirer's response to modernity is superior to that of the members of the Frankfurt School (see Chapter Three). Though Cornell and Panfilio are sympathetic to the critique of modernity that Horkheimer, Adorno, and Marcuse present, they agree with Habermas's worry that this critique gives way to pessimism. However, they argue that it is ultimately not Habermas's solution, which privileges the philosophy of language and communicative rationality, but rather a Cassirer-inspired solution, which emphasizes pluralism and the transformative power of symbolic forms, that is the best response to the problems of modernity. Indeed, on their view, acknowledging the pluralism of the symbolic forms undermines the hegemony of modernity insofar as it reveals that, "instrumental rationality is . . . merely one symbolic form of the world among so many others" (Cornell and Panfilio 2010: 93). And by coupling this pluralism with a recognition of the revolutionary capacity the symbolic forms have to transform humanity, they argue that Cassirer's view points toward how this plurality of symbolic forms can ultimately reconfigure humanity in a way that undermines the hegemony of modernity, capitalism, and colonialism.

In order to highlight the revolutionary dimensions of Cassirer's position, Cornell and Panfilio bring it to bear on the issues of colonialism and anti-Black racism and the revolutionary response they call for. In so doing, they highlight the affinities between Cassirer's view and the theory of revolution offered by Fanon and Henry, according to which the revolution required is not just a matter of the acquisition of political power, but requires a "reconstituted humanity" (Cornell and Panfilio 2010: 95). Drawing out the parallels between Cassirer's view and Henry's view, in particular, they argue that Cassirer's recognition of the staying power of myth dovetails in important ways with Henry's emphasis on the need for engagement with Afro-Caribbean thought and for a creolized dialogue between its historicist-socialist and poeticist schools, which can serve as the means to a symbolic reconfiguration of a new humanity beyond the domination of anti-Black racism (see Cornell and Panfilio 2010: Ch. 4).[42]

Though Cornell and Panfilio thus emphasize Cassirer's value for critical theory from a different perspective than does Habermas, they all agree that it is a mistake to regard Cassirer's philosophy of symbolic forms as an outmoded humanism. To the contrary, they argue that his views embody a still viable form of humanism, which has much to offer in the revolutionary efforts to resist the forces of capitalism, colonialism, and racism by reconfiguring humanity in a new, symbolic way.

Summary

My goal in this chapter was to bring to light the wide-ranging legacy of Cassirer's wide-ranging philosophical system. I began by analyzing the impact of Cassirer's philosophy of science on contemporary debates in philosophy of science about structural realism and the dynamic *a priori*, before turning to his impact on social science in the work of Goldstein and Lewin. I then considered the influence of Cassirer's historical writings on the development of intellectual history in the 20th century. At that point, I shifted my focus to the contribution Cassirer's philosophy of symbolic forms made to the semantic turn in 20th century thought. In this vein, I started with a discussion of the reception of Cassirer's philosophy of symbolism in general by the philosophers, Langer, Goodman, and Merleau-Ponty. I next explored the uptake of his view in more specific areas, including art history and the iconographical method of Panofsky; Blumenberg's study of myth; and the theories of language defended by members of the Bakhtin circle. I concluded with an analysis of Cassirer's impact on critical theory and Habermas's, Cornell's, and Panfilio's account of the value Cassirer's humanism still has today. It is in these myriad ways that Cassirer's philosophy of the whole human person, as a being who lives and breathes in the cultural world she helps create, continues to resonate.

Notes

1 Worrall (1989) is credited with introducing this idea in contemporary debates.
2 See Gower (2000): §3, Massimi (2011): §1.2.2, and Heis (2014b): §3.
3 See French and Ladyman (2003): §3, Cei and French (2009), French (2014): Ch. 4, §§8–12.
4 See Heis (2014b): §3 and Mormann (2015): §6, who also argues that OSR's move to eliminate objects completely is at odds with Cassirer's view, and that Cassirer's structuralism is "more comprehensive" insofar as he extends it not just to quantum physics, as OSR tends to do, but to classical and non-classical physics as well (59).
5 See Massimi (2011) for the suggestion that Cassirer's view lends itself to semantic structuralism, according to which structural realism fixes the good epistemic conditions under which we can be justified in making assertions about unobservable objects (§1.4). See Mormann (2015): 59–61 for the suggestion that Cassirer's view can be read as supporting van Fraassen's empiricist structuralism, according to which, "science represents the empirical phenomena as embeddable in certain *abstract structures*," which are "describable only up to structural isomorphism" (60).
6 See Stump (2011) for an overview of Pap's theory of the functional *a priori*.
7 For a discussion of the relationship between Cassirer's and Reichenbach's theories of the *a priori*, see Richardson (1998): Ch. 5, Ryckman (2005): Ch. 2, and Heis (2013).
8 See Chapter Four for further discussion and complete references.
9 Although I shall not discuss this here, Cassirer's writings on psychology have also been influential for S.H. Foulkes's group theoretical approach to psychology and Hanscarl Leuner's symbol-based approach to LSD research. See Nitzgen (2010) for a discussion of the relationship between Goldstein, Cassirer, and Foulkes.
10 See Andersch (2015) for a discussion of Cassirer's influence on pre- and postwar developments of psychopathology.
11 Goldstein also later approvingly cites Cassirer's conception of 'life' and 'mind' not as two distinct substances, but rather as two modes of 'forming' and 'acting' (1995: 356–359).
12 See Heis (2013) for a discussion of Cassirer's influence on Lewin.

13 See Friedman (2005a), (2005b), (2008) for a discussion of Cassirer's influence on the history of science and Kuhn.

14 I discuss the significance of Cassirer's historical writings for Bakhtin below.

15 Cassirer and Kristeller, a Renaissance historian, were colleagues at Hamburg in the 1920s and they stayed in close contact while they were both in exile. It is Kristeller's work The Philosophy of Marsilio Ficino (1943 [1964]) that most bears the marks of his engagement with Cassirer.

16 Foucault praises Cassirer's synthetic approach to the Enlightenment in his review of the French translation of Cassirer's Enlightenment book, "Une histoire restée muette" (1967 [2001]). See Moynahan (2013): xxiii–xxvi for a discussion of Foucault's reading of Cassirer.

17 While Langer had been familiar with Cassirer's Philosophy of Symbolic Forms in the 1930s, they met for the first time in 1942 in the U.S. and as correspondence between them reveals they had a mutually productive intellectual relationship until Cassirer's death.

18 She draws out the parallels between her view of symbols and Cassirer's in "On a New Definition of 'Symbol'" in Philosophical Sketches (1956 [1962]).

19 While I will take up Langer's discussion of art below, see Schultz (2000) for a discussion of the relationship between Cassirer's and Langer's approach to myth.

20 For Langer's discussion of intuitive and imaginative modes of understanding, see, e.g., Feeling and Form, 1953: 28–29, 40–41, 378–379 and Problems of Art, 1957: Ch. 5.

21 See Pollok (2016) for a discussion of the parallels between Cassirer's and Langer's aesthetics. See Makkreel (2020) for a discussion of the relationship between the aesthetics of Cassirer, Langer, and Dilthey.

22 In Problems of Art, Langer gestures towards how her view might relate to art that represents nature, arguing that it would involve a "subjectification of nature" in which we make nature "itself a symbol of life and feeling" (1957: 73). In subordinating even representational art to expression, her view thus seems to part ways from Cassirer who recognizes the intuitive forms of external objects as distinct from the intuitive forms of human emotions.

23 See Guyer (2014): Volume 3, Ch. 9 and 10.4 and Carter (2015) for a discussion of the affinities between the aesthetics of Cassirer, Langer, and Goodman.

24 See Friedman (2000) and Gordon (2010) for a discussion of the legacy of the Davos disputation.

25 Though I shall not pursue it here, Cassirer's philosophy also intersects with Husserl's in various ways. For example, in Cassirer's discussion of how logical essences come to consciousness through mental acts in Substance and Function, he approvingly cites Husserl's discussion of the relationship between logic and phenomenology in Logical Investigations (SF 24–25). And in the second volume of The Philosophy of Symbolic Forms, Cassirer praises Husserl both for developing an anti-psychologistic approach to the phenomenology and for the application of this phenomenology to different regions within culture (PSFv2 32 fn. 15/14 fn. 12/12, fn. 7). Cassirer also offers a more critical discussion of Husserl's account

of the noetic and hyletic layers of consciousness as involving a problematic conception of the relationship between form and content in his chapter on symbolic pregnance in the third volume (see PSFv3 231–237/223–228/196–200). For a discussion of the relationship between Cassirer's philosophy and Husserl's phenomenology, see Martell (2015) and de Warren (2015).

26 In the *Phenomenology*, Merleau-Ponty's references to *The Philosophy of Symbolic Forms* primarily occur in the context of his discussion of pathology (see (2013): 126 fn. 60–62, 129 fn. 67, 197–198 fn. 30), the body and expression (see 244 fn. 67); and space (see 303–304 fnn. 88, 92). See Matherne (2014) for an analysis of the influence of Cassirer's account of the imagination and pathology (PSFv3, Pt. 2, Ch. 6) on Merleau-Ponty's account of pathology and human existence. See Endres (2016) for a discussion of the ways in which Cassirer anticipates Merleau-Ponty's phenomenology of perception.

27 For other references to symbolic pregnance, see Merleau-Ponty (2013): 154, 453, (1964): 12.

28 See Ferretti (1989) and Levine (2013) for a discussion of the relationship between Cassirer, Panofsky, and Warburg. See Holly (1984): Chapter 5 and Alloa (2015) for a discussion of the relationship between the aesthetics of Cassirer and Panofsky. See Moynahan (2005) for a discussion of the political dimensions of their relationship in the context of the Davos Debate.

29 In addition to Cassirer's influence on Panofsky's framework for iconography, the two collaborated together in Hamburg. In 1924, for example, they delivered a two-part lecture series on the concept of 'idea' in the history of aesthetics: whereas Cassirer delivered the first lecture on the concept of the 'idea' in Plato's aesthetics ("*Eidos and Eidolon*: The Problem of Beauty and Art in the Dialogues of Plato"), Panofsky delivered the 'sequel' in which he traces the fate of the 'idea' in aesthetics from antiquity to the 17th century, and which he develops into *Idea: A Concept in Art Theory* (1968).

30 The essay was originally printed in 1924–1925 as part of the *Vorträge der Bibliothek Warburg*, but was then printed separately in 1927 (translation 1991).

31 See, for example, "The Problem of Style in the Visual Art" ("*Das Problem des Stils in der bildenden Kunst*") (1915) and "The Concept of Artistic Volition" ("*Der Begriff des Kunstwollens*") (1921 [1981]).

32 Panofsky here appears to have in mind Cassirer's 1923 definition of a symbolic form at CSF 76/79.

33 For a comparison of Cassirer's and Lévi-Strauss's symbolic approach to myth, see Silverstone (1976).

34 See Rudolph (2003), Pavesich (2008), Barash (2011), and Haverkamp (2016) for a discussion of the relationship between Cassirer and Blumenberg.

35 Translations are my own.

36 In "Affinities and Dominances" ("*Affinitäten und Dominanzen*") published posthumously in *Ein mögliches Selbstverständnis* (1997), Blumenberg compares the Davos disputation to Luther and Zwingli's debate over the Eucharist, with Luther

and Heidegger defending a substance-based view, and Zwingli and Cassirer presenting a function-based one.

37 Cassirer's philosophy of language was also influential for the American philosopher of language Wilbur Marshall Urban. See Urban's "Cassirer's Philosophy of Language" (1949) for discussion of Cassirer's significance for his project and the claim that Cassirer was the "first . . . to develop a philosophy of language in the full sense of the word," i.e., a systematic account of the nature and function of language (Urban 1949: 403).

38 See Brandist (1997), (2002): Chs. 5–6 and Poole (1998) for a discussion of Cassirer's influence on Bakhtin and the Bakhtin Circle.

39 More recently, Jennifer Marra has been exploring a Cassirer-style account of comedy. See Marra (2015) for an account of comedy as a symbolic form and (2018) for a discussion of the politics and morality of humor.

40 See Cornell and Panfilio (2010): Ch. 3 for a discussion of the similarities and dissimilarities between Cassirer and Habermas.

41 In this work, Cornell builds on her earlier reading of Cassirer's value for critical theory in *Moral Images of Freedom* (2008).

42 See Cornell and Panfilio (2010): Ch. 5 and Conclusion for an application of this conception of revolution to the dialectic between anti-Black racism and revolution in South Africa.

Further reading

In addition to the readings referenced in footnotes throughout this chapter, see also:

Barash, Jeffrey Andrew, ed. (2008). *The Symbolic Construction of Reality: The Legacy of Ernst Cassirer*. Chicago: The University of Chicago Press. A collection of essays on Cassirer's legacy.

Schlipp, Paul Arthur, ed. (1949). *The Library of Living Philosophers, Volume VI: The Philosophy of Ernst Cassirer*. Evanston: Library of Living Philosophers. Originally intended as a series of collected essay on Cassirer's philosophy to which Cassirer would respond; however, Cassirer died before he was able to do so.

Glossary

absolute *a priori* the kind of *a priori* that remains the same across all phenomena or experience.

analogical phase a phase in which a symbolic form encourages an active understanding of subjects as agents and of cultural configurations as analogues of a reality that we produce through our agency.

anti-psychologism a philosophical approach that rejects any attempt to account for the conditions of experience, science, or culture exclusively in terms of psychic acts, operations, or processes that occur in individual minds.

appearance (Erscheinung) an object that depends on the forms or functions of the mind. Also known as a 'phenomenon'.

categories the *a priori* concepts, functions, or relations that articulate the conditions that make objects and experience possible.

cognition (Erkenntnis) an intellectual mode of grasping ideal functions or relations (concepts, laws, and principles).

configuration (Gestaltung) a particular cultural object or practice, which has its source in a symbolic form.

constitutive *a priori* the kind of *a priori* that constitutes objects of cognition.

constitutive concept a concept that articulates the conditions that make the objects of cognition possible.

constructive method a method that clarifies the forms, functions, or relations that make possible the objective forms of culture.

Copernican Revolution the revolution Kant calls for in metaphysics, which requires rejecting the idea that cognition conforms to objects and embracing the alternative idea that objects conform to cognition.

copy theory (Abbildtheorie) of cognition a theory, according to which cognition is a process in which the mind copies mind-independent substances. Also translated as 'picture theory'.

critical idealism a theory, according to which the object of cognition is not a mind-independent substance ('thing in itself', 'noumenon'), but rather an 'appearance' or 'phenomenon' that conforms to the forms or functions of the mind.

critical theory of cognition a theory of cognition as a process in which objects of cognition conform to the forms or functions of the mind.

critical philosophy philosophy done in the spirit of Kant.

cultural sciences (Kulturwissenschaften) the set of empirical sciences that study human cultural phenomena, e.g., the study of religion, art, language, history, etc., in what is known today as the humanities and (some) social sciences. Also known as 'human sciences' (Geisteswissenschaften).

experience (Erfahrung) often used in the technical Marburg way to refer to mathematical-scientific cognition as a whole. Sometimes used in a looser way to refer to lived-experience.

expressive perception (Ausdruckswahrnehmung) the kind of perception grounded in the function of expression, in which we perceive things in affective or subjective terms.

function-concept (Funktionsbegriff) a concept that serves as a function, i.e., a relation that orders elements into a series and can be modeled as $\emptyset$ (a, b, c...). Also referred to as a 'relation-concept' (Relationsbegriff).

function of expression (Ausdrucksfunktion) the specific function of consciousness that involves representing things in subjective or affective terms in expressive perception.

function of presentation (Darstellungsfunktion) the specific function of consciousness that involves representing things in more objective terms in intuition.

function of pure signification (reine Bedeutungsfunktion) the specific function that involves representing purely ideal relations in cognition.

function of representation (Repräsentation) the function that characterizes consciousness in general and that involves representing one content of consciousness in and through another.

human sciences (Geisteswissenschaften) the set of empirical sciences that study human cultural phenomena, e.g., the study of religion, art, language, history, etc., in what is known today as the humanities and (some) social sciences. See 'cultural sciences'.

intuition (Anschauung) a mode of spatial and/or temporal representation, which represents things in objective terms.

invariant relation a relation or function that remains the same across phenomena, experience, or culture.

lived-experience (Erlebnis) the conscious experience had by an individual.

logical structuralism a theory about mathematics, according to which mathematics is the science of ideal structures that are constructed on the basis of logical functions or relations.

logic of relations the part of modern symbolic logic concerned with the study of logical relations or functions.

mimetic phase a phase in which a symbolic form encourages a passive conception of subjects and of cultural configurations as passive copies of reality

natural right (Naturrecht) a right one has in virtue of being a human being.

natural sciences (Naturwissenschaften) the empirical sciences that study physical phenomena, like physics and chemistry.

objective spirit (Geist) spirit as it manifests in the collective, historically unfolding effort human beings engage in to build a shared cultural world.

original contract the social contract understood as an ideal principle of the legitimacy of legislation.

perception (Wahrnehmung) a mode of sensible representation, which represents things either in affective terms in 'expressive perception' or in more objective terms in 'thing perception' or 'intuition'.

philosophical anthropology a philosophical tradition that focuses on questions about the nature of the human being.

physical statement (*Aussage*) a statement in natural science that expresses empirical invariant relations.

positivism an approach to philosophy that rejects *a priori* methodology in favor of an empirical methodology in which philosophy is led by the natural sciences.

principle of simplicity a regulative principle that demands that natural science progress towards theories that comprehend the greatest number of phenomena with the fewest possible concepts, laws, and principles.

principle of systematicity a regulative principle that demands that natural science progress towards theories that are continuous and unitary.

psychologism an account of the conditions of the possibility of experience, science, or culture exclusively in terms of psychic acts, operations, or processes that occur in individual minds.

reconstructive method a method by means of which the psychic acts, processes, and operations involved in consciousness and lived-experience are reconstructed in critical thought.

regulative *a priori* the kind of *a priori* that regulates the progress of scientific inquiry.

regulative principle a principle that regulates or guides scientific inquiry.

relativized *a priori* the kind of *a priori* that varies from scientific theory to scientific theory.

republican constitution a constitution, which recognizes the autonomy, equality, and inalienable rights of human beings.

sensuous form in art, a medium-specific form, like line or color in painting, rhythm or rhyme in poetry.

statements of laws (*Gesetzesaussagen*) a specific kind of physical statement in natural science that refers to the physical laws that govern general classes of physical phenomena.

statements of measurement (*Massaussagen*) a specific kind of physical statement in natural science that expresses the facts that pertain to the measurement of individual physical phenomena.

statements of principles (*Prinzipienaussagen*) a specific kind of physical statement in natural science that refers to the universal physical principles that govern all physical phenomena.

structuralism in philosophy of mathematics, an ontological theory, according to which mathematical objects are not defined in terms of any intrinsic properties, but rather in terms of their place in ideal structures.

structural realism in philosophy of science, the theory that our current best scientific theories are approximately true accounts not of unobservable objects, but rather of mathematical structures.

substance-concept (Substanzbegriff) a concept that is a copy of an absolutely mind-independent object.

symbolic animal (animal symbolicum) the definition of the human being as an animal who engages in symbolic activities that produce a cultural world.

symbolic form a region of culture defined as an active forma formans that generates cultural configurations and provides a unique context of signification in the cultural world.

symbolic phase a phase in which a symbolic form encourages a recognition of subjects as spontaneous and autonomous and of cultural configurations as the product of this spontaneity and autonomy.

symbolic pregnance (Prägnanz) a description of the relationship between a symbol and what it symbolizes, such that a symbol is full of and imprinted by what it symbolizes.

technique of myth a strategy utilized by fascist governments that involves reintroducing certain myths through technological means in order to promote a more mythical mode of thinking.

thing in itself (Ding an sich) defined in a metaphysical sense as an absolute substance that is entirely independent from the mind. Defined in a regulative sense as the goal of completeness and systematicity that scientific inquiry progresses towards. Also known as a 'noumenon'.

thrownness (Geworfenheit) a Heideggerean concept that refers to the way in which human beings are thrown into a particular temporal situation.

transcendental logic a logic that studies the a priori conditions of the possibility of the cognition of objects.

transcendental method the philosophical method endorsed by
the Marburg Neo-Kantians, which requires orienting oneself
around the 'facts' of science and culture, and offering an account
of the conditions that make those facts possible.

transcendental statement (*Aussage*) a statement made in critical
philosophy that expresses the *a priori* invariant relations that
make scientific experience possible.

Bibliography

Works by Cassirer

References listed here are to the **original**; the corresponding edition in **ECW** (*Gesammelte Werke: Hamburger Ausgabe*, edited by Birgit Recki, Hamburg: Meiner) or in **ECN** (*Nachgelassene Manuskripte und Texte*, Hamburg: Meiner); and the **English translation**.

(1899) *Descartes' Kritik der mathematischen und naturwissenschaftlichen Erkenntnis* (printed in *Leibniz' System*). (1998) ECW 1.

(1902) *Leibniz' System in seinen wissenschaftlichen Grundlagen.* Marburg: N.G. Elwert. (1998) ECW 1.

(1906) *Das Erkenntnisproblem in der Philosophie und Wissenschaft der neueren Zeit.* Bd. 1. Berlin: Bruno Cassirer. (1999) ECW 2.

(1907) *Das Erkenntnisproblem in der Philosophie und Wissenschaft der neueren Zeit.* Bd. 2. Berlin: Bruno Cassirer. (1999) ECW 3.

(1907) "Kant und die moderne Mathematik." *Kant-Studien* 12: 1–40. (2001) ECW 9, 37–82.

(1910) *Substanzbegriff und Funktionsbegriff: Untersuchungen über die Grundfragen der Erkenntniskritik.* Berlin: Bruno Cassirer. (2000) ECW 6. Translated as (1923) *Substance and Function and Einstein's Theory of Relativity.* Translated by William Swabey and Marie Swabey. Chicago: Open Court.

(1912) "Hermann Cohen und die Erneurung der Kantischen Philosophie." *Kant-Studien* 17: 252–273. (2001) ECW 9, 119–138. Translated as (2015) "Hermann Cohen and the Renewal of Kantian Philosophy." Translated by Lydia Patton. In *The Neo-Kantian Reader*, edited by Sebastian Luft, 221–235. London: Routledge.

(1916) *Freiheit und Form. Studien zur deutschen Geistesgeschichte.* Berlin: Bruno Cassirer. (2001) ECW 7.

(1918) *Kants Leben und Lehre.* Supplementary Volume to *Immanuel Kants Werke.* Berlin: Bruno Cassirer. (2001) ECW 8. Translated as (1981) *Kant's Life and Thought.* Translated by James Haden. New Haven: Yale University Press.

(1920) *Das Erkenntnisproblem in der Philosophie und Wissenschaft der neueren Zeit. Die nachkantischen System*, Bd. 3. Berlin: Bruno Cassirer. (2000) ECW 4.

(1921) *Zur Einsteinschen Relativitätstheorie. Erkenntnistheoretische Betrachtungen.* Berlin: Bruno Cassirer. (2001) ECW 10. Translated as (1923) *Substance and Function and Einstein's Theory of Relativity.* See above.

(1921) *Idee und Gestalt. Goethe, Schiller, Hölderlin, Kleist.* Berlin: Bruno Cassirer. (2001) ECW 9, 243–348.

(1923) *Philosophie der symbolischen Formen. Erster Teil: Die Sprache.* Berlin: Bruno Cassirer. (2001) ECW 11. Translated as (2021) *The Philosophy of Symbolic Forms, Volume One: Language.* Translated by Steve G. Lofts. London and New York: Routledge. Also translated as (1955) *The Philosophy of Symbolic Forms. Volume One: Language.* Translated by Ralph Manheim. New Haven: Yale University Press.

(1923) "Der Begriff der Symbolischen Form im Aufbau der Geisteswissenschaften." In *Vorträge der Bibliothek Warburg*: Vorträge 1921–1922, 11–39. Leipzig: B.G. Teubner. (2003) ECW 16, 75–104. Translated as (2013) "The Concept of Symbolic Form in the Construction of the Human Sciences" in *The Warburg Years*, 72–100.

(1923) "Die Kantischen Elemente in Wilhelm von Humboldts Sprachphilosophie." *Festschrift für Paul Hensel*, 105–127. Greiz: Ohag. (2003) ECW 16, 105–134. Translated as (2013) "The Kantian Elements in Wilhelm von Humboldt's Philosophy of Language." In *The Warburg Years*, 101–129.

(1924) "Eidos und Eidolon: Das Problem das Schönen und der Kunst in Platons Dialogen." In *Vorträge der Bibliothek Warburg*: Vorträge 1922–1923, 1–27. Leipzig: B.G. Teubner. (2003) ECW 16, 135–164. Translated as "Eidos and Eidolon: The Problem of Beauty and Art in the Dialogues of Plato." In *The Warburg Years*, 214–243.

(1925) *Philosophie der symbolischen Formen. Zweiter Teil: Das mythische Denken.* Berlin: Bruno Cassirer. (2002) ECW 12. Translated as (2021) *The Philosophy of Symbolic Forms. Volume Two: Mythical Thinking.* Translated by Steve G. Lofts. London and New York: Routledge. Also translated as (1955) *The Philosophy of Symbolic Forms. Volume Two: Mythical Thought.* Translated by Ralph Manheim. New Haven: Yale University Press.

(1925) "Sprache und Mythos: Ein Beitrag zum Problem der Götternamen." *Studien der Bibliothek Warburg* VI. Leipzig: B.G. Teubner. (2003) ECW 16, 227–312. Translated as (2013) "Language and Myth: A Contribution to the Problem of the Names of the Gods" in *The Warburg Years*, 130–213. Also translated as (1946) *Language and Myth.* Translated by Susanne K. Langer. New York: Harper.

(1925) "Paul Natorp." *Kant-Studien* 30: 273–298. (2003) ECW 16, 197–226.

(1927) *Individuum und Kosmos in der Philosophie der Renaissance. Studien der Bibliothek Warburg* X. Leipzig: B.G. Teubner. (2002) ECW 14, 1–222. Translated as (1963) *The Individual and the Cosmos in Renaissance Philosophy.* Translated by Mario Domandi. New York: Harper & Row.

(1927) "Das Symbolproblem und seine Stellung im System der Philosophie." *Zeitschrift für Ästhetik und allgemeine Kunstwissenschaft* 21: 191–208. (2004) ECW 17, 253–282. Translated as (2013) "The Problem of the Symbol and Its Place in the System of Philosophy." In *The Warburg Years*, 254–271.

(1928) "Die Idee der Republikanischen Verfassung: Rede zuf Verfassungsfeier am 11. August 1928." Hamburg: Friederischen. (2004) ECW 17, 291–307. Translated as (2018) "The Idea of a Republican Constitution," Translated by Seth Berk, *Philosophical Forum* 49(1), 3–17.

(1928) *"Zur Metaphysik der symbolischen Formen."* In (1995) ECN 1, Hrsg. v. John Michael Krois, 3–109. Translated as (1996) "On the Metaphysics of Symbolic Forms" in *The Philosophy of Symbolic Forms: Volume 4: The Metaphysics of Symbolic Forms.* Translated by John Michael Krois. New Haven: Yale University Press, 3–111.

(c. 1928) *"Beilage: Symbolbegriff: Metaphysik des Symbolischen."* (1995) ECN 1, Hrsg. v. John Michael Krois, 261–274. Translated as (1996) "Appendix: The Concept of the Symbol: Metaphysics of the Symbolic (c. 1921–27)." Translated by John Michael Krois. New Haven: Yale University Press, 223–234.

(1929) *Philosophie der symbolischen Formen. Dritter Teil: Phänomenologie der Erkenntnis.* Berlin: Bruno Cassirer. (2002) ECW 13. Translated as (2021) *The Philosophy of Symbolic Forms, Volume Three: The Phenomenology of Cognition.* Translated by Steve G. Lofts. London and New York: Routledge. Also translated as (1957) *The Philosophy of Symbolic Forms. Volume Three: The Phenomenology of Knowledge.* Translated by Ralph Manheim. New Haven: Yale University Press.

(1929) "Neo-Kantianism." In *Encyclopedia Britannica*, 14th edition, edited by J. L. Garvin and Franklin Henry Hooper. London: Encyclopedia Britannica. (2004) ECW 17, 308–315.

(1930) "Form und Technik." In *Kunst und Technik*, edited by Leo Kestenberg. Berlin: Volksverband der Bücherfreude-Verband. (2004) ECW 17, 139–184. Translated as "Form and Technology." In *The Warburg Years*, 272–316.

(1930) "'Geist' und 'Leben' in der Philosophie der Gegenwart." *Die Neue Rundschau* 41: 244–64. (2004) ECW 17, 185–206. Translated as (2020) "Appendix: 'Spirit' and 'Life' in Contemporary Philosophy" in *The Philosophy of Symbolic Forms, Volume Three: The Phenomenology of Cognition.* Translated by Steve G. Lofts. Abingdon: Routledge, 561–583.

(1930) "Wandlungen der Staatsgesinnung und der Staatstheorie in der deutschen Geschichte." (2011) ECN 7, Hrsg. v. Jörn Bohr, Gerald Hartung, 85–102.

(1931) "Mythischer, ästhetischer und theoretischer Raum." *Zeitschrift für Ästhetik und allgemeine Kunstwissenschaft* 25, Beiheft: *Vierter Kongreß für Ästhetik und allgemeine Kunstwissenschaft*, 21–36. (2004) ECW 17, 411–432. Translated as "Mythical, Aesthetic, and Theoretical Space." In *The Warburg Years*, 317–333.

(1931) "Kant und das Problem der Metaphysik. Bemerkungen zu Martin Heideggers Kantinterpretation." *Kant-Studien* 36: 1–26. (2004) ECW 17, 221–52.

(1932) *Die Platonische Renaissance in England und die Schule von Cambridge.* Studien der Bibliothek Warburg 24. Leipzig: B.G. Teubner. (2002) ECW 14, 223–380. Translated as (1953) *The Platonic Renaissance in England.* Translated by James Pettegrove. New York: Gordian Press.

(1932) *Die Philosophie der Aufklärung.* Tübingen: Mohr. (2003) ECW 15. Translated as (1951) *The Philosophy of the Enlightenment.* Translated by Fritz Koelln and J. Pettegrove. Princeton: Princeton University Press.

(1932) *Goethe und die GeschichtlicheWelt: Drei Aufsätze.* Berlin: Bruno Cassirer. (2004) ECW 18, 355–436.

(1932) "Das Problem Jean-Jacques Rousseau." *Archiv für Geschichte der Philosophie* 41, 177–213, 479–513. (2004) ECW 18, 3–82. Translated as (1954) *The Question of Jean-Jacques Rousseau.* Translated by Peter Gay. New York: Columbia University Press.

(1932) "Die Sprache und der Aufbau der Gegenstandswelt." *Bericht über den XII. Kongress der deutschen Gesellschaft für Psychologie.* Jena: G. Fischer. (2004) ECW 18, 111–126 (German version), 265–90 (French Version: "*Le langage et la construction du monde des objects*")]. Translated as (2013) "Language and the Construction of the World of Objects." In *TheWarburgYears,* 334–362.

(1932) "Vom Wesen und Werden des Naturrechts." *Zeitschrift für Rechtsphilosophie* 6(1). Leipzig: Meiner. (2004) ECW 18, 203–228.

(1933) "Das Problem Jean Jacques Rousseau" *Archiv für Geschichte der Philosophie* 41. (2004) ECW 18, 3–82. Translated as (1989) *The Question of Jean Jacques Rousseau.* Translated by Peter Gay. New Haven: Yale University Press.

(1933) Cassirer, Ernst to Küchler, Walther 27/04/1933. Letter in the Warburg Archive. WIA, GC [General Correspondence], E. Cassirer to W. Küchler, 27 April 1933.

(1935) "*Der Begriff der Philosophie als Problem der Philosophie.*" (2008) ECN 9, Hrsg. v. John Michael Krois and Christian Möckel, 141–165. Translated as (1979) "The Concept of Philosophy as a Philosophical Problem." Translated by Donald Phillip Verne. In *Symbol, Myth, and Culture,* 49–63.

(1936) *Determinismus und Indeterminismus in der modernen Physik.* Göteborgs Högskolas Arsskrift 42. ECW 19. Translated as (1955) *Determinism and Indeterminism in Modern Physics.* Translated by O. Theodor Benfey. New Haven: Yale University Press.

(1936) "Critical Idealism as a Philosophy of Culture." In *Symbol, Myth, and Culture,* 64–91. (2011) ECN 7, Hrsg. v. Jörn Bohr, Gerald Hartung, 93–119.

(1938) "Le Concept de Groupe et la Théorie de la Perception." *Journal de Psychologie:* 368–414. (2007) ECW 24. Translated as "The Concept of Group and the Theory of Perception." Translated by Aron Gurwitsch. *Philosophy and Phenomenological Research* 5(1): 1–36. ECW 24, 209–250.

(1939) *Descartes. Lehre—Personlichkeit—Wirkung.* Stockholm: Bermann-Fischer Verlag. (2005) ECW 20.

(1939) *Axel Hägerström: Eine Studie zur Schwedischen Philosophie der Gegenwart.* Göteborg: Göteborgs Högskolas Årsskrift 45. (2005) ECW 21, 3–118.

(1939) "Was ist 'Subjektivismus'?" *Theoria* 5: 111–140. (2006) ECW 22, 167–192.

(1939) "Thorilds Stellung in der Geistesgeschichte des 18. Jahrhunderts." *Svenska Historie—Vitterhetens-och Antikvitet-Akademiens Handlingar,* 1941. (2005) ECW 21, 119–236.

(1940) "Descartes und die Königin Christina von Schweden." (2005) ECW 20, 128–204.

(1941) "Thorild und Heder." *Theoria* 7: 75–92. (2007) ECW 24, 37–52.

(1941-2) "Descartes, Leibniz, and Vico." In *Symbol, Myth, and Culture,* 95–107.

(1942) *Zur Logik der Kulturwissenschaften: Fünf Studien.* Göteborg: Göteborgs Högskolas Årsskrift 47. (2007) ECW 24, 357–490. Translated as (2000) *The Logic of the*

Cultural Sciences: Five Studies. Translated by S.G. Lofts. New Haven: Yale University Press. Translated earlier as (1961) *The Logic of the Humanities*. Translated by Clarence Smith Howe. New Haven: Yale University Press.

(1942) "Hegel's Theory of the State." In *Symbol, Myth, and Culture*, 108–120.

(1942) "Language and Art" I and II. In *Symbol, Myth, and Culture*, 145–195. (2011) ECN 7, Hrsg. v. Jörn Bohr, Gerald Hartung, 141–184.

(1942) "The Philosophy of History." In *Symbol, Myth, and Culture*, 121–141.

(1943) "The Educational Value of Art." In *Symbol, Myth and Culture*, 196–215. (2011) ECN 7, Hrsg. v. Jörn Bohr, Gerald Hartung, 185–200.

(1943) "Hermann Cohen, 1842–1918." *Social Research* 10(2): 218–232. (2007) ECW 24, 161–174.

(1944) *An Essay on Man: An Introduction to a Philosophy of Human Culture*. New Haven: Yale University Press. (2006) ECW 23.

(1944) "Judaism and the Modern Political Myths." *Contemporary Jewish Record* 7: 115–126. (2007) ECW 24: 197–208.

(1944) "Philosophy and Politics" in *Symbol, Myth, and Culture*, 219–232.

(1945) "The Technique of Our Modern Political Myths" in *Symbol, Myth, and Culture*, 242–267.

(1945) *Rousseau—Kant—Goethe*. The History of Ideas Series, Vol. 1, Princeton. (2007) ECW 24: 491–578. Translated by James Gutmann, Paul Oscar Kristeller and John Herman Randall, Jr. The History of Ideas Series, Volume 1. Princeton: Princeton University Press.

(1946) *The Myth of the State*. New Haven: Yale University Press. (2007) ECW 25.

(1946) "Albert Schweitzer as Critic of Nineteenth-Century Ethics." *The Albert Schweitzer Jubilee Book*. Edited by A. A. Roback. Cambridge: Sci-Art Publishers, 239–258. (2007) ECW 24, 321–334.

(1957) *Das Erkenntnisproblem in der Philosophie und Wissenschaft der neueren Zeit. Vierter Band: Von Hegels Tod bis zur Gegenwart*. (1832–1932). (2000) ECW 5. Translated as (1950) *The Problem of Knowledge: Philosophy, Science, and History Since Hegel*. Translated by William H. Woglom and Charles W. Hendel. New Haven: Yale University Press, 1969.

(1979) *Symbol, Myth, and Culture: Essays and Lectures of Ernst Cassirer: 1935–1945*. Edited by Donald Phillip Verene. New Haven: Yale University Press.

(2009) *Ausgewählter wissenschaftlicher Briefwechsel*. Edited by John Michael Krois. Meiner: Hamburg. (2009) ECN 18, Hrsg. v. John Michael Krois.

(2013) *The Warburg Years (1919–1933): Essays on Language, Art, Myth, and Technology*. Translated by S.G. Lofts and A. Calcagno. New Haven: Yale University Press.

***

Other works

Alloa, Emmanuel. (2015) "Could Perspective Ever Be a Symbolic Form?" *Journal of Aesthetics and Phenomenology* 2(1): 51–71.

Andersch, Norbert. (2015) "Symbolic Form and Mental Illness: Ernst Cassirer's Contribution to a New Concept of Psychopathology." In *The Philosophy of Ernst Cassirer: A Novel Assessment*, edited by J. Tyler Friedman and Sebastian Luft, 163–198. Berlin: De Gruyter.

Anderson, Lanier. (2005) "Neo-Kantianism and the Roots of Anti-Psychologism." *British Journal for the History of Philosophy* 13(2): 287–323.

Bakhtin, Mikhail. (1984) *Rabelais and His World*. Bloomington: Indiana University Press.

———(2007). *The Bildungsroman and Its Significance in the History of Realism*. Toronto: Parasitic Ventures Press.

Barash, Jeffrey Andrew, ed. (2008) *The Symbolic Construction of Reality: The Legacy of Ernst Cassirer*. Chicago: University of Chicago Press.

———(2011) "Myth in History, Philosophy of History as Myth: On the Ambivalence of Hans Blumenberg's Interpretation of Ernst Cassirer's Theory of Myth." *History and Theory* 50(3): 328–340.

Baumann, Charlotte. (2019) "Hermann Cohen on Kant, Sensations, and Nature in Science." *Journal of the History of Philosophy* 57(4): 647–674.

Bayer, Thora Ilin. (2001) *Cassirer's Metaphysics of Symbolic Forms: A Philosophical Commentary*. New Haven: Yale University Press.

Beiser, Fred. (2014) *The Genesis of Neo-Kantianism, 1796–1880*. Oxford: Oxford University Press.

Biagioli, Francesca. (2015) "Cassirer's View of the Mathematical Method as a Paradigm of Symbolic Thinking." *Lectiones & Acroases Philosophicae* 8(1): 193–223.

———(2016) *Space, Number, and Geometry from Helmholtz to Cassirer*. Dordrecht: Springer.

———(2020). "Ernst Cassirer's transcendental account of mathematical reasoning." *Studies in History and Philosophy of Science* 79: 30–40.

Blumenberg, Hans. (1974) "Ernst Cassirers gedenkend." *Revue Internationale de Philosophie* 28 (110 (4)): 456–63.

———(1985) *Work on Myth*. Cambridge: MIT Press.

———(1997) *Ein mögliches Selbstverständnis*. Stuttgart: Phillipp Reclam.

———(2001) "Anthropologische Annäherung an die Aktualität der Rhetorik." In *Ästhetische und metaphorologische Schriften*, edited by Anselm Haverkamp. Suhrkamp.

Bottici, Chiara. (2007) *A Philosophy of Political Myth*. New York: Cambridge University Press.

Brandist, Craig. (1997) "Bakhtin, Cassirer and Symbolic Forms." *Radical Philosophy* 85: 20–27.

———(2002) *The Bakhtin Circle: Philosophy, Culture and Politics*. London: Pluto Press.

Bundgaard, Peer F. (2011) "The Grammar of Aesthetic Intuition: On Ernst Cassirer's Concept of Symbolic Form in the Visual Arts." *Synthese* 179(1): 43–57.

Capeillères, Fabien. (2007) "Philosophy as Science: 'Function' and 'Energy' in Cassirer's 'Complex System' of Symbolic Forms." *The Review of Metaphysics* 61(2): 317–377.

————(2008) "History and Philosophy in Ernst Cassirer's System of Symbolic Forms." In *The Symbolic Construction of Reality*, edited by Jeffrey Andrew Barash, 40–72. Chicago: University of Chicago Press.

————(2012/13) "Artistic Intuition Within Cassirer's System of Symbolic Forms. A Brief Reassessment." *Cassirer Studies* V/VI, 91–124.

————(2015) "Cassirer on the 'Objectivity' of Evil: The Symbolic Constitution of Der Mythus des 20 Jahrhunderts." In *The Philosophy of Ernst Cassirer: A Novel Assessment*, edited by J. Tyler Friedman and Sebastian Luft, 435–468. Berlin: De Gruyter.

Carter, Curtis L. (2015) "After Cassirer: Art and Aesthetic Symbols in Langer and Goodman." In *The Philosophy of Ernst Cassirer: A Novel Assessment*, edited by J. Tyler Friedman and Sebastian Luft, 401–418. Berlin: De Gruyter.

Carus, A. W. (2007) *Carnap and Twentieth-Century Thought: Explication as Enlightenment.* Cambridge: Cambridge University Press.

Cassirer Studies (2010) *Cassirer Studies* Volume III: *The Originary Phenomena.* http://www.cassirerstudies.eu/?cat=43

Cassirer, Toni Bondy. (2003) *Mein Leben mit Ernst Cassirer.* Hamburg: Meiner.

Cei, Angelo, and Steven French. (2009) "On the Transposition of the Substantial into the Functional: Bringing Cassirer's Philosophy of Quantum Mechanics into the Twenty-First Century." In *Constituting Objectivity*, 95–115. The Western Ontario Series in Philosophy of Science. Dordrecht: Springer.

Clarke, Evan. (2019). "Neo-Kantianism." In *A Companion to Nineteenth-Century Philosophy*, edited by John Shand, 389–417. Hoboken: Blackwell.

Cohen, Hermann. (1871/1885) *Kants Theorie der Erfahrung.* Berlin: F. Dümmler.

————(1877) *Kants Begründung der Ethik.* Berlin: F. Dümmler.

————(1889) *Kants Begründung der Aesthetik.* Berlin: F. Dümmler.

————(1902/1922) *System der Philosophie. 1: Logik der reinen Erkenntnis.* Berlin: Bruno Cassirer.

————(1904/1907) *System der Philosophie 2: Ethik des Reinen Willens.* Berlin: Bruno Cassirer.

————(1912) *System der Philosophie. 3:Ästhetik des reinen Gefühls.* Berlin: Bruno Cassirer.

————(1915) *Der Begriffe der Religion im System der Philosophie.* Geissen: Töpelmann.

————(1919). *Die Religion derVernunft aus den Quellen des Judentums.* Leipzig: Fock.

Cornell, Drucilla. (2008) *Moral Images of Freedom: A Future for Critical Theory.* Lanham: Rowan & Littlefield Publishers.

Cornell, Drucilla, and Kenneth Michael Panfilio. (2010) *Symbolic Forms for a New Humanity: Cultural and Racial Reconfigurations of Critical Theory.* New York: Fordham University Press.

Coskun, Deniz. (2007) *Law as Symbolic Form: Ernst Cassirer and the AnthropocentricView of Law.* Law and Philosophy Library 82. Dordrecht: Springer.

Curthoys, Ned. (2013) *The Legacy of Liberal Judaism: Ernst Cassirer and Hannah Arendt's Hidden Conversation.* New York: Berghahn Books.

Dahlstrom, Daniel. (2021) "Cassirer's Phenomenological Affinities." In *Interpreting Cassirer: Critical Essays*, edited by Simon Truwant, 193–213. Cambridge: Cambridge University Press.

Dedekind, Richard. (1888). *Was sind und was sollen die Zahlen?* Braunschweig: Vieweg.

de Warren, Nicolas. (2015) *"Reise um die Welt:* Cassirer's cosmological phenomenology." In *New Approaches to Neo-Kantianism*, edited by Nicolas de Warren and Andrea Staiti, 82–108. Cambridge: Cambridge University Press

———(2021) "Spirit in the Age of Technical Production." In *Interpreting Cassirer: Critical Essays*, edited by Simon Truwant, 109–129. Cambridge: Cambridge University Press.

Edgar, Scott. (2008) "Paul Natorp and the Emergence of Anti-Psychologism in the Nineteenth Century." *Studies in History and Philosophy of Science* 39(1): 54–65.

———(2020) "Hermann Cohen," *The Stanford Encyclopedia of Philosophy* (Winter 2020 Edition), Edward N. Zalta (ed.), forthcoming URL = <https://plato.stanford.edu/archives/win2020/entries/cohen/>.

Eilenberger, Wolfram. (2020) *Time of the Magicians: Wittgenstein, Benjamin, Cassirer, Heidegger, and the Decade that Reinvented Philosophy*. Translated by Shaun Whiteside. New York: Penguin.

Endres, Tobias. (2016) "Die Philosophie der symbolischen Formen als Phänomenologie der Wahrnehmung." In *Philosophie der Kultur- und Wissensformen: Ernst Cassirer neu lessen*, edited by Tobias Endres, Pellegrino Favuzzi, and Timo Klattenhoff, 35–54. Frankfurt: Peter Lang.

———(2020) *Ernst Cassirers Phänomenologie der Wahrnehmung*. Hamburg: Felix Meiner Verlag.

Ferrari, Massimo. (2002). "Sources for the History of the Concept of Symbol from Leibniz to Cassirer." In *Symbol and Physical Knowledge*, edited by Massimo Ferrari and Ion-Olipiu Stamatescu, 3–32. Berlin: Springer.

———(2003). *Ernst Cassirer. Stationen einer philosophischen Biographie*. Translated by Marion Lauschke. Hamburg: Meiner. Translation of (1996). *Cassirer. Dalla Scuola di Marburgo alla filosofia della cultura*. Florence: Olschki.

———(2009) "Is Cassirer a Neo-Kantian Methodologically Speaking?" In *Neo-Kantianism in Contemporary Philosophy*, edited by Rudolf A. Makkreel and Sebastian Luft. Bloomington: Indiana University Press.

———(2012) "Between Cassirer and Kuhn. Some Remarks on Friedman's Relativized a Priori." *Studies in History and Philosophy of Science* 43 (1): 18–26.

———(2015) "Cassirer and the Philosophy of Science." In *New Approaches to Neo-Kantianism*, edited by Nicolas de Warren and Andrea Staiti, 261–284. Cambridge: Cambridge University Press.

———(2021) "Science As a Symbolic Form: Ernst Cassirer's Culture of Reason." In *Interpreting Cassirer: Critical Essays*, edited by Simon Truwant, 72–88. Cambridge: Cambridge University Press.

Ferretti, Silvia. (1989) *Cassirer, Panofsky, and Warburg: Symbol, Art, and History*. Translated by Richard Pierce. New Haven: Yale University Press.

Foucault, Michel. (2001) "Une Histoire Restée Muette." In *Dits et Écrits: 1954–1988*. Quarto. Paris: Gallimard.

French, Steven. (2014) *The Structure of the World: Metaphysics and Representation*. Oxford: Oxford University Press.

French, Steven, and James Ladyman. (2003) "Remodelling Structural Realism: Quantum Physics and the Metaphysics of Structure." *Synthese* 136(1): 31–56.

Freudenthal, G. (2008) "The Hero of the Enlightenment," in *The Symbolic Construction of Reality: The Legacy of Ernst Cassirer*, edited by J. Barash, 189–214. Chicago: The University of Chicago Press.

Friedman, J. Tyler and Luft, Sebastian, eds. (2015). *The Philosophy of Ernst Cassirer: A Novel Assessment*. Berlin: De Gruyter.

Friedman, Michael. (2000) *A Parting of the Ways: Carnap, Cassirer, and Heidegger*. Chicago: Open Court.

———(2005a) "Ernst Cassirer and the Philosophy of Science." In *Continental Philosophy of Science*, edited by Gary Gutting, 69–83. Malden: Blackwell.

———(2005b) "Ernst Cassirer and Contemporary Philosophy of Science." *Angelaki*, 10(1): 119–128.

———(2008) "Ernst Cassirer and Thomas Kuhn: The Neo-Kantian Tradition in History and Philosophy of Science." *The Philosophical Forum*, Special Issue: Classical Neo-Kantianism, edited by Andrew Chignell, Terence Irwin, and Thomas Teufel, 39(2): 239–252.

Gawronsky, Dimitry. (1949) "Ernst Cassirer: His Life and his Work." In *Library of Living Philosophers: The Philosophy of Ernst Cassirer*, edited by Paul Schlipp, 1–38. Evanston: Library of Living Philosophers.

Gay, Peter. (1967) "The Social History of Ideas: Ernst Cassirer and After." In *The Critical Spirit*, edited by Herbert Marcuse, Kurt H. Wolff, and Barrington Moore, 106–120. Boston: Beacon Press.

Goldstein, Kurt. (1995) *The Organism: A Holistic Approach to Biology Derived from Pathological Data in Man*. New York: Zone Books.

Goodman, Nelson. (1976) *Languages of Art: An Approach to a Theory of Symbols*. Second edition. Indianapolis: Hackett.

———(1978) *Ways of Worldmaking*. Indianapolis: Hackett.

Gordon, Peter Eli. (2010) *Continental Divide: Heidegger, Cassirer, Davos*. Cambridge: Harvard University Press.

———(2012) "German Idealism and German Liberalism in the 1920s: The Case of Ernst Cassirer." In *The Weimar Moment: Liberalism, Political Theology, and Law*, edited by Leonard Kaplan and Rudy Koshar, 337–344. Lanham: Lexington Books.

Gower, Barry. (2000) "Cassirer, Schlick and 'Structural' Realism: The Philosophy of the Exact Sciences in the Background to Early Logical Empiricism." *British Journal for the History of Philosophy* 8(1): 71–106.

Guyer, Paul. (2014) *A History of Modern Aesthetics*. Cambridge: Cambridge University Press.

Habermas, Jürgen. (2001) "The Liberating Power of Symbols: Ernst Cassirer's Humanistic Legacy and the Warburg Library." In *The Liberating Power of Symbols: Philosophical Essays*. Translated by Peter Dews, 1–29. Cambridge: MIT Press.

Hansson, Jonas, and Svante Nordin. (2006) *Ernst Cassirer: The Swedish Years*. Bern: Peter Lang.

Haverkamp, Anselm. (2016) "Blumenberg in Davos: The Cassirer—Heidegger Controversy Reconsidered." *MLN* 131(3): 738–753.

Heidegger, Martin. (1962). *Being and Time*. Translated by John Macquarrie and Edward Robinson. San Francisco: Harper & Row.

———(1990) *Kant and the Problem of Metaphysics*. Translated by Richard Taft. Fourth edition. Studies in Continental Thought. Bloomington: Indiana University Press.

Heis, Jeremy. (2010) "'Critical Philosophy Begins at the Very Point Where Logistic Leaves Off': Cassirer's Response to Frege and Russell." *Perspectives on Science* 18(4): 383–408.

———(2011) "Ernst Cassirer's Neo-Kantian Philosophy of Geometry." *British Journal for the History of Philosophy* 19(4): 759–794.

———(2013) "Ernst Cassirer, Kurt Lewin, and Hans Reichenbach." In *The Berlin Group and the Philosophy of Logical Empiricism*, 67–94. Boston Studies in the Philosophy and History of Science. Dordrecht: Springer.

———(2014a) "Ernst Cassirer's Substanzbegriff und Funktionsbegriff." *HOPOS: The Journal of the International Society for the History of Philosophy of Science* 4(2): 241–270.

———(2014b) "Realism, Functions, and the a Priori: Ernst Cassirer's Philosophy of Science." *Studies in History and Philosophy of Science* 48: 10–19.

———(2015) "Arithmetic and Number in the Philosophy of Symbolic Forms." In *The Philosophy of Ernst Cassirer: A Novel Assessment*, edited by J. Tyler Friedman and Sebastian Luft, 123–140. Berlin: De Gruyter.

———(2018). "Neo-Kantianism." *The Stanford Encyclopedia of Philosophy*, edited by Edward N. Zalta. URL = <https://plato.stanford.edu/archives/sum2018/entries/neo-kantianism/>.

Hendel, Charles. (1949) "Ernst Cassirer." In *Library of Living Philosophers: The Philosophy of Ernst Cassirer*, edited by Paul Schlipp, 55–59. Evanston: Library of Living Philosophers.

Hilbert, David. (1899) "Grundlagen der Geometrie." First edition. In *Festschrift zur Feier der Enthüllung des Gauss-Weber-Denkmals in Göttingen*, 1–92. Leipzig: Teubner.

Hoel, Aud Sissel, and Ingvild Folkvord, eds. (2012) *Ernst Cassirer on Form and Technology: Contemporary Readings*. Houndmills: Palgrave Macmillan.

Holly, Michael Ann. (1984) *Panofsky and the Foundations of Art History*. Ithaca: Cornell University Press.

Holzhey, Helmut. (2005) "Cohen and the Marburg School in Context." In *Hermann Cohen's Critical Idealism*, edited by Reiner Munk, 3–37. Dordrecht: Springer.

Ihmig, Karol-Nobert. (1997) *Cassirers Invariantentheorie der Erfahrung und seine Rezeption des 'Erlangers Programms'*. Hamburg: Meiner.

————(1999) "Ernst Cassirer and the Structural Conception of Objects in Modern Science: The Importance of the 'Erlanger Program.'" *Science in Context* 12(4): 513–529.

Kant, Immanuel. (1996) "On the Common Saying: That may be correct in theory, but it is of no use in practice." In *Practical Philosophy*. Translated by Mary Gregor, 273–310. Cambridge: Cambridge University Press.

————"Toward Perpetual Peace." In *Practical Philosophy*. Translated by Mary Gregor, 311–352. Cambridge: Cambridge University Press.

————*The Metaphysics of Morals*. In *Practical Philosophy*. Translated by Mary Gregor, 353–604. Cambridge: Cambridge University Press.

————(1997) *Prolegomena to Any Future Metaphysics*. Translated by Gary Hatfield. Cambridge: Cambridge University Press. [Prol.]

————(1998) *Critique of Pure Reason*. Translated by Paul Guyer and Allen W. Wood. Cambridge: Cambridge University Press. [CPR]

————(1902) *Kants gesammelte Schriften*. Deutschen (formerly, Königlichen Preussichen) Akademie der Wissenschaften (Eds.). Berlin: Walter de Gruyter.

Keller, Pierre. (2015). "Cassirer's Retrieval of Kant's Copernican Revolution in Semiotics." In *The Philosophy of Ernst Cassirer: A Novel Assessment*, edited by J. Tyler Friedman and Sebastian Luft, 259–288. Berlin: De Gruyter.

Köhnke, Klaus Christian. (1991) *The Rise of Neo-Kantianism: German Academic Philosophy between Idealism and Positivism*. Cambridge: Cambridge University Press.

Kreis, Guido. (2010) *Cassirer und die Formen des Geistes*. Berlin: Suhrkamp.

————(2015) "The Varieties of Perception Non-Conceptual Content in Kant, Cassirer, and McDowell." In *The Philosophy of Ernst Cassirer: A Novel Assessment*, edited by J. Tyler Friedman and Sebastian Luft, 313–338. Berlin: De Gruyter.

————(2021) "Cassirer's Philosophy of Mind: From Consciousness to "Objective Spirit"." In *Interpreting Cassirer: Critical Essays*, edited by Simon Truwant, 170–190. Cambridge: Cambridge University Press.

Krijnen, Christian and Noras, Andrzej, eds. (2012) *Marburg versus Südwestdeutschland: philosophische Differenzen zwischen den beiden Hauptschulen des Neukantianismus*. Würzburg: Königshausen & Neumann.

Kristeller, Paul Oskar. (1964). *The Philosophy of Marsilio Ficino*. Translated by Virginia Lanphear Conant. Gloucester, MA: Peter Smith.

Krois, John Michael. (1987) *Cassirer: Symbolic Forms and History*. New Haven: Yale University Press.

Langer, Susanne K. (1930). *The Practice of Philosophy*. New York: H. Holt and Company.

————(1937) *An Introduction to Symbolic Logic*. Boston: Houghton Mifflin Company.

————(1949) "On Cassirer's Theory of Language and Myth." In *The Library of Living Philosophers: The Philosophy of Ernst Cassirer*, edited by Paul Arthur Schlipp. Evanston: Library of Living Philosophers.

————(1953) *Feeling and Form*. New York: Charles Scribner's Sons.

———(1957a) Philosophy in a New Key: A Study in the Symbolism of Reason, Rite, and Art. Third edition. Cambridge: Harvard University Press.

———(1957b) Problems of Art: Ten Philosophical Lectures. London: Routledge & K. Paul.

———(1962) Philosophical Sketches. Baltimore: Johns Hopkins Press.

———(1967), (1972), (1982) Mind: An Essay on Human Feeling. Vols. 1–3. Baltimore: Johns Hopkins Press.

Leib, Robert. (2021) "Interaction between Language and the Other Symbolic Forms." In Interpreting Cassirer: Critical Essays, edited by Simon Truwant, 13–33. Cambridge: Cambridge University Press.

Leinkauf, Thomas, ed. (2003) Dilthey und Cassirer : die Deutung der Neuzeit als Muster von Geistes- und Kulturgeschichte. Hamburg: Meiner.

Levine, Emily J. (2013) Dreamland of Humanists: Warburg, Cassirer, Panofsky, and the Hamburg School. Chicago: The University of Chicago Press.

Lewin, Kurt. (1931) "The Conflict Between Aristotelian and Galileian Modes of Thought in Contemporary Psychology." The Journal of General Psychology 5(2): 141–177.

———(1949) "Cassirer's Philosophy of Science and the Social Science." In Library of Living Philosophers: The Philosophy of Ernst Cassirer, edited by Paul Arthur Schlipp. Evanston: Library of Living Philosophers.

Liebmann, Otto. (1865). Kant und die Epigonen: Eine Kritische Abhandlung. Stuttgart: Carl Schober.

Lofts, S.G. (2000) Ernst Cassirer: A "Repetition" of Modernity. Albany: State University of New York Press.

Lorde, Audre. (1997) The Collected Poems of Audre Lorde. First edition. New York: W.W. Norton.

Luft, Sebastian. (2015) The Space of Culture: Towards a Neo-Kantian Philosophy of Culture (Cohen, Natorp, and Cassirer). Oxford: Oxford University Press.

———ed. (2015) The Neo-Kantian Reader. London: Routledge.

———(2021). "Cassirer's Place in Today's Philosophical Landscape: "Synthetic Philosophy," Transcendental Idealism, Cultural Pluralism." In Interpreting Cassirer: Critical Essays, edited by Simon Truwant, 214–236. Cambridge: Cambridge University Press.

Makkreel, Rudolf A. (2020) "Cassirer, Langer, and Dilthey on the Distinctive Kinds of Symbolism in the Arts." Journal of Transcendental Philosophy, https://doi.org/10.1515/jtph-2019-0023.

Makkreel, Rudolf A., and Sebastian Luft, eds. (2010) Neo-Kantianism in Contemporary Philosophy. Bloomington: Indiana University Press.

Mali, Joseph. (2008) "Myth of the State Revisited: Ernst Cassirer and Modern Political Theory—Chicago Scholarship." In The Symbolic Construction of Reality: The Legacy of Ernst Cassirer, edited by Jeffrey Andrew Barash, 135–162. Chicago: University of Chicago Press.

Marra, Jennifer. (2015) "Humor as a Symbolic Form: Cassirer and the Culture of Comedy." In The Philosophy of Ernst Cassirer: A Novel Assessment, edited by J. Tyler Friedman and Sebastian Luft, 419–434. Berlin: De Gruyter.

————(2018) "Make Comedy Matter: Ernst Cassirer on the Politics and Morality of Humour." *European Journal of Humour Research* 6(1): 162–171.

Marrow, Alfred J. (1969) *The Practical Theorist: The Life and Work of Kurt Lewin.* New York: Basic Books.

Martell, Timothy. (2015) "Cassirer and Husserl on the Phenomenology of Perception." *Studia Phaenomenologica* 15: 413–431.

Massimi, Michela. (2011) "Structural Realism: A Neo-Kantian Perspective." In *Scientific Structuralism*, edited by Alisa Bokulich and Peter Bokulich, 1–23. Dordrecht: Springer.

Matherne, Samantha. (2014) "The Kantian Roots of Merleau-Ponty's Account of Pathology." *British Journal for the History of Philosophy* 22(1): 124–149.

————(2015) "Marburg Neo-Kantianism as Philosophy of Culture." In *The Philosophy of Ernst Cassirer: A Novel Assessment*, edited by J. Tyler Friedman and Sebastian Luft, 201–232. Berlin: De Gruyter.

————(2018) "Cassirer's Psychology of Relations: From the Psychology of Mathematics and Natural Science to the Psychology of Culture." *Journal of the History of Analytic Philosophy*, Special Issue: Method, Science, and Mathematics: Neo-Kantianism and Early Analytic Philosophy, edited by Lydia Patton and Scott Edgar, 6(3): 132–162.

————(2021) "The Status of Art in Cassirer's System of Culture." In *Interpreting Cassirer: Critical Essays*, edited by Simon Truwant, 34–52. Cambridge: Cambridge University Press.

McLear Colin. (2020) "Kantian Conceptualism/Nonconceptualism," *The Stanford Encyclopedia of Philosophy* (Summer 2020 Edition), Edward N. Zalta (ed.), URL = <https://plato.stanford.edu/archives/sum2020/entries/kant-conceptualism/>.

Meland, Ingmar. (2012/13) "The Doctrine of Basis Phenomena. A Phenomenological Foundation for the Philosophy of Symbolic Forms?" *Cassirer Studies* V/VI, 31–63.

Merleau-Ponty, Maurice. (1951) *The Prose of the World.* Translated by John O'Neill. Evanston: Northwestern University Press.

———— (1964) *The Primacy of Perception.* Translated by William Cobb. Evanston: Northwestern University Press.

————(2013) *Phenomenology of Perception.* Translated by Donald Landes. New York: Routledge.

Mormann, Thomas. (2008) "Idealization in Cassirer's Philosophy of Mathematics." *Philosophia Mathematica* 16(2): 151–181.

————(2015) "From Mathematics to Quantum Mechanics: On the Conceptual Unity of Cassirer's Philosophy of Science." In *The Philosophy of Ernst Cassirer: A Novel Assessment*, edited by J. Tyler Friedman and Sebastian Luft, 31–64. Berlin: De Gruyter.

Moss, Gregory S. (2015) *Ernst Cassirer and the Autonomy of Language.* Lanham: Lexington Books.

Moynahan, Gregory B. (2013) *Ernst Cassirer and the Critical Science of Germany, 1899–1919.* London: Anthem Press.

————(2005). "The Davos Debate, Science, and the Violence of Interpretation: Panofsky, Heidegger, and Cassirer on the Politics of History." In *Exile, Science and Bildung: The Contested Legacies of German Emigre Intellectuals*, edited by David Kettler and Gerhard Lauer, 111–124. New York: Palgrave.

Munk, Reiner. (2005) *Hermann Cohen's Critical Idealism*. Dordrecht: Springer.

Natorp, Paul. (1888) *Einleitung in die Psychologie nach kritischer Methode*. Mohr.

————(1912) *Allgemeine Psychologie nach kritischer Methode*. J. C. B. Mohr (P. Siebeck).

————(1918) "Hermann Cohens philosophische leistung unter dem gesichtspunkte des systems." *Philosophische Vorträge veröffentlicht von der Kant-Gesellschaft*, 21, edited by Arthur Liebert. Berlin: Reuther & Reichard.

————(1921) *Die logischen Grundlagen der exakten Wissenschaften*, Second edition. Leipzig: Teubner.

————(2015a) "Kant and the Marburg School." In *The Neo-Kantian Reader*, edited by Sebastian Luft, translated by Francesca Bottenberg, 180–195. London: Routledge.

————(2015b) "On the Objective and Subjective Grounding of Knowledge." In *The Neo-Kantian Reader*, edited by Sebastian Luft, translated by Lois Phillips and David Kolb, 164–179. London: Rouledge.

Naumann, Barbara and Recki, Birgit, eds. (2002). *Cassirer und Goethe: Neue Aspekte einer philosophisch-literarischen Wahlverwandtschaft. Studien aus dem Warburg-Haus 5*. Berlin: Akademie Verlag.

Nitzgen, Dieter. (2010) "Hidden Legacies. S.H. Foulkes, Kurt Goldstein and Ernst Cassirer." *Group Analysis* 43(3): 354–371.

Panofsky, Erwin. (1915) "Das Problem des Stils in der Bildenden Kunst." *Zeitschrift für Ästhetik und Allgemeine Kunstwissenschaft* 10: 460–467.

————(1968) *Idea: A Concept in Art Theory*. Translated by Joseph Peake. Columbia: University of South Carolina Press.

————(1972) *Studies in Iconology ; Humanistic Themes in the Art of the Renaissance*. New York: Harper & Row.

————(1981) "The Concept of Artistic Volition." Translated by Kenneth J. Northcott and Joel Snyder. *Critical Inquiry* 8(1): 17–33.

————(1991) *Perspective as Symbolic Form*. Translated by Christopher Wood. New York: Zone Books.

————(2008). "On the Relationship of Art History and Art Theory: Towards the Possibility of a Fundamental System of Concepts for a Science of Art." Translated by Katharina Lorenz and Jas' Elsner. *Critical Inquiry* 35: 43–71.

Pap, Arthur. (1943a) "On the Meaning of Necessity." *The Journal of Philosophy* 40(17): 449–458.

————(1943b) "On the Meaning of Universality." *The Journal of Philosophy* 40(19): 505–514.

————(1944) "The Different Kinds of A Priori." *The Philosophical Review* 53: 465–484.

————(1946) *The A Priori in Physical Theory*. New York: Russell & Russell.

Pavesich, Vida. (2008) "Hans Blumenberg's Philosophical Anthropology: After Heidegger and Cassirer." *Journal of the History of Philosophy* 46(3): 421–48.

Plümacher, Martina. (2021) "Rethinking Representation: Cassirer's Philosophy of Human Perceiving, Thinking, and Understanding." In *Interpreting Cassirer: Critical Essays*, edited by Simon Truwant, 151–189. Cambridge: Cambridge University Press.

Pollok, Anne. (2015) "The First and Second Person Perspective in History: Or, Why History Is 'Culture Fiction.'" In *The Philosophy of Ernst Cassirer: A Novel Assessment*, edited by J. Tyler Friedman and Sebastian Luft, 341–360. Berlin: De Gruyter.

————(2016) "Significant Formation: An Intersubjective Approach to Aesthetic Experience in Cassirer and Langer." *Graduate Faculty Philosophy Journal* 37(1): 71–95.

————(2021) "History As an Expression and Interpretation of Human Culture." In *Interpreting Cassirer: Critical Essays*, edited by Simon Truwant, 53–71. Cambridge: Cambridge University Press.

Poole, Brian. (1998) "Bakhtin and Cassirer: The Philosophical Origins of Carnival Messianism." *The South Atlantic Quarterly* 97(3): 537–578.

Reck, Erich (2020). "Cassirer's Reception of Dedekind and the Structuralist Transformation of Mathematics." In *The Pre-History of Mathematical Structuralism*, edited by Erich Reck and Georg Schiemer. Oxford: Oxford University Press.

Reck, Erich, and Pierre Keller. (forthcoming) "From Dedekind to Cassirer: Logicism and the Kantian Heritage." In *Kant's Philosophy of Mathematics, Vol. II: Reception and Development After Kant*, edited by Ofra Rechter and Carl Posy. Cambridge: Cambridge University Press.

Recki, Birgit (2003). *Kultur als Praxis: Eine Einführung in die Philosophie Ernst Cassirers.* Berlin: De Gruyter.

————, ed. (2012) *Philosophie der Kultur—Kultur des Philosophierens: Ernst Cassirer im 20. und 21. Jahrhundert.* Hamburg: Meiner.

Reichenbach, Hans. (1965). *The Theory of Relativity and A Priori Knowledge*, translated by Maria Reichenbach. Berkeley: University of California Press.

Renz, Ursula. (2002) *Die Rationalität der Kultur : Zur Kulturphilosophie und ihrer Transzendentalen Begründung bei Cohen, Natorp und Cassirer.* Hamburg: Felix Meiner.

————(2011) "From Philosophy to Criticism of Myth: Cassirer's Concept of Myth." *Synthese* 179(1): 135–152.

————(2020) "Cassirer's Enlightenment: On Philosophy and the 'Denkform' of Reason." *British Journal for the History of Philosophy* 28(3): 636–652.

Richardson, Alan. (1998) *Carnap's Construction of the World: The Aufbau and the Emergence of Logical Empiricism.* Cambridge; New York: Cambridge University Press.

————(2003) "Conceiving, Experiencing, and Conceiving Experiencing: Neo-Kantianism and the History of the Concept of Experience." *Topoi* 22(1): 55–67.

————(2006) "'The Fact of Science' and Critique of Knowledge: Exact Science as Problem and Resource in Marburg Neo-Kantianism." In *The Kantian Legacy* in *Nineteenth-century Science*, edited by Michael Friedman and Alfred Nordmann, 221–226. Cambridge, MA: MIT Press.

———(2016) "Cassirer's Substance Concept and Functional Concept." In *Ten Neglected Classics of Philosophy*, edited by Eric Schliesser, 177–194. Oxford: Oxford University Press.

Rickert, Heinrich. (1926) *Kulturwissenschaft und Naturwissenschaft*. Tubingen: J. C. B. Mohr (P. Siebeck).

———(1986) *The Limits of Concept Formation in Natural Science*. Cambridge: Cambridge University Press.

Rudolph, Enno, ed. (1999) *Cassirers Weg zur Philosophie der Politik*, Cassirer-Forschungen 5. Hamburg: Meiner.

———(2003) *Ernst Cassirer im Kontext*. Tübingen: Mohr Siebeck.

———(2008) "Symbol and History: Ernst Cassirer's Critique of the Philosophy of History." In *The Symbolic Construction of Reality: The Legacy of Ernst Cassirer*, edited by Jeffrey Andrew Barash, 3–16. Chicago: University of Chicago Press.

Ryckman, Thomas. (2005) *The Reign of Relativity.: Philosophy in Physics 1915–1925*. Oxford: Oxford University Press.

———(2015) "A Retrospective View of Determinism and Indeterminism in Modern Physics." In *The Philosophy of Ernst Cassirer: A Novel Assessment*, edited by J. Tyler Friedman and Sebastian Luft, 65–102. Berlin: De Gruyter.

———(2018) "Cassirer and Dirac on the Symbolic Method in Quantum Mechanics: A Confluence of Opposites." *Journal of the History of Analytic Philosophy* Special Issue: Method, Science, and Mathematics: Neo-Kantianism and Early Analytic Philosophy, edited by Lydia Patton and Scott Edgar, 6(3): 213–243.

———(2021) "Quantum Mechanics As the Ultimate Mode of Symbol Formation: The Final Stage of Cassirer's Philosophy of Physical Science." In *Interpreting Cassirer: Critical Essays*, edited by Simon Truwant, 89–108. Cambridge: Cambridge University Press.

Saxl, Fritz. (1949) "Ernst Cassirer." In *Library of Living Philosophers: The Philosophy of Ernst Cassirer*, edited by Paul Schlipp, 47–51. Evanston: Library of Living Philosophers.

Schiemer, Georg. (2018) "Cassirer and the Structural Turn in Modern Geometry." *Journal of the History of Analytic Philosophy*, Special Issue: Method, Science, and Mathematics: Neo-Kantianism and Early Analytic Philosophy, edited by Lydia Patton and Scott Edgar, 6(3): 183–212.

Schliesser, Eric, ed. (2016) *Ten Neglected Classics of Philosophy*. New York: Oxford University Press.

Schlipp, Paul Arthur, ed. (1949) *Library of Living Philosophers: The Philosophy of Ernst Cassirer*. Evanston: Library of Living Philosophers.

Schmitz-Rigal, Christiane. (2011) "Science and Art: Physics as a Symbolic Formation." *Synthese* 179(1): 21–41.

Schultz, William. (2000) *Cassirer and Langer on Myth: An Introduction*. Theorists of Myth, vol. 12. New York: Garland.

Seidengart, Jean. (2012) "Cassirer, Reader, Publisher, and Interpreter of Leibniz's Philosophy." In *New Essays on Leibniz Reception*, edited by Ralf Krömer and Yannick Chin-Drian, 129–142. Basel: Springer.

Silverstone, Roger. (1976) "Ernst Cassirer and Claude Lévi-Strauss: Two Approaches to the Study of Myth." *Archives de Sciences Sociales des Religions* 21(41): 25–36.

Skidelsky, Edward. (2008) *Ernst Cassirer: The Last Philosopher of Culture*. Princeton: Princeton University Press.

Spengler, Oswald. (1991 [1918]) *The Decline of the West*. Ed. Arthur Helps, and Helmut Werner. Trans. Charles F. Atkinson. New York: Oxford University Press.

Strauss, Leo. (1947) "Review of The Myth of the State by Ernst Cassirer." *Social Research* 14(1): 125–128.

———(1988) *What Is Political Philosophy?: And Other Studies*. Chicago: University of Chicago Press.

Stump, David J. (2011) "Arthur Pap's Functional Theory of the A Priori." HOPOS: *The Journal of the International Society for the History of Philosophy of Science* 1(2): 273–290.

Truwant, Simon. (2014) "Cassirer's Enlightened View on the Hierarchy of the Symbolic Forms and the Task of Philosophy." *Cassirer Studies VII–VIII: Darstellung/Vorstellung*: 119–139.

———(2021) ed., *Interpreting Cassirer: Critical Essays*. Cambridge: Cambridge University Press.

———(2021) "Political Myth and the Problem of Orientation: Reading Cassirer in Times of Cultural Crisis." In *Interpreting Cassirer: Critical Essays*, edited by Simon Truwant, 130–150. Cambridge: Cambridge University Press.

Urban, Wilbur Marshall. (1949) "Cassirer's Philosophy of Language." In *Library of Living Philosophers: The Philosophy of Ernst Cassirer*, edited by Paul Arthur Schlipp. Evanston: Library of Living Philosophers.

Verene, Donald Phillip. (2011) *The Origins of the Philosophy of Symbolic Forms: Kant, Hegel, and Cassirer*. Evanston: Northwestern University Press.

Voloshinov, V. N. (1986) *Marxism and the Philosophy of Language*. Translated by Ladislav Matejka and I.R. Titunik. Cambridge, MA: Harvard University Press.

———(2004) "Report on Work as a Postgraduate Student." In *The Bakhtin Circle: In the Master's Absence*, edited by David Shepherd, Craig Brandist, and Galin Tihanov. Manchester: Manchester University Press.

Willey, Thomas E. (1978) *Back to Kant: The Revival of Kantianism in German Social and Historical Thought, 1860–1914*. Detroit: Wayne State University Press.

Windelband, Wilhelm. (1980) "History and Natural Science." *History and Theory* 19(2): 165–168.

Worrall, John. (1989) "Structural Realism: The Best of Both Worlds?" *Dialectica* 43(1–2): 99–124.

Yap, Audrey. (2017) "Dedekind and Cassirer on Mathematical Concept Formation." *Philosophia Mathematica* 25(3): 369–389.

Index